AN INTRODUCTION TO FOURIER ANALYSIS

An Introduction to Fourier Analysis

R. D. STUART
M.A., Ph.D., A.M.I.E.E.
Professor of Electrical Engineering
Northeastern University
Boston, Mass., U.S.A.

LONDON
CHAPMAN AND HALL

First published 1961
by Methuen & Co Ltd
Reprinted 1965

First published as a Science Paperback 1966
by Chapman and Hall Ltd
11 *New Fetter Lane, London EC4P 4EE*
Reprinted 1969, 1974, 1977

ISBN 0 412 20200 *X*

Printed in Great Britain
at the University Press, Cambridge

Distributed in the USA
by Halsted Press
a division of John Wiley & Sons, Inc.
New York

Contents

Preface

Fourier Analysis has now become a widely used technique in many branches of science and engineering. This book has been written mainly with the needs of students of these subjects in mind. Thus full mathematical rigour has not been attempted. On the other hand an attempt has been made to avoid presenting the subject in such a way that it becomes a mere 'tool'. An elementary proof of the fundamental theorems has been given to provide some insight into the more theoretical aspects of the subject and to provide a better basis for the understanding of some of the phenomena which occur later, such as the Gibbs' phenomenon. The first two chapters deal with Fourier Series. These are followed by two on Fourier Integrals and finally two on applications. It is common to think of Fourier Analysis in terms of the analysis of time varying functions. The final chapter shows that this point of view is unnecessarily restricted and that the theory is of much wider applicability.

In compiling a work of this nature the work of many other authors has of course been drawn upon and in particular H. S. Carslaw, *Theory of Fourier Series and Integrals* should be mentioned. Other works are acknowledged in the text. The author's views on the subject were influenced to a considerable extent also by a series of lectures given by Mr J. A. Ratcliffe at the Cavendish Laboratory in 1948. Finally, the author would like to thank Professor R. D. Klein for reading the manuscript and making many helpful criticisms and suggestions.

CHAPTER I

Fourier Series

1.1 Introduction

Our studies in acoustics introduce us to the idea of harmonics. A string fixed at each end can vibrate at a number of different frequencies. The lowest of these, f, is called the fundamental frequency. The others will have values $2f$, $3f$, etc., and are called harmonics. These frequencies can be excited simultaneously and the resulting vibration has a complex waveform. This will still be periodic and the period T is equal to $(1/f)$. Thus we see that the summation of a number of frequencies harmonically related gives rise to a complex periodic waveform. Conversely a complex periodic waveform can be analysed into a number of sinusoidal variations which are harmonically related. It is the second process we shall be concerned with in this chapter. The plot of amplitude against frequency which in this case consists of a number of discrete lines at frequencies $f, 2f, \ldots$ is called the spectrum of the waveform.

1.2 The Fourier Series

Suppose we have a function $f(t)$ which is periodic and suppose for simplicity that the period is 2π, i.e. $f(t+2\pi)=f(t)$. Then $f(t)$ can be represented as a series

$$\begin{aligned} f(t) &= \tfrac{1}{2}a_0+a_1\cos t+a_2\cos 2t+\ldots+b_1\sin t+b_2\sin 2t+\ldots \\ &= \tfrac{1}{2}a_0+\sum_{n=1}^{\infty}(a_n\cos nt+b_n\sin nt) \end{aligned}$$

This is the Fourier series representation of the function $f(t)$. As might be expected certain restrictions must be placed on $f(t)$ for the expression to be valid. The conditions which must be satisfied, known as Dirichlet's conditions, are met by practically all of the functions which are of interest in the physical sciences. The integral $\int_{-\pi}^{\pi}|f(t)|\,dt$ must be finite and $f(t)$ must be piecewise continuous and piecewise monotonic. If $f(t)$ is piecewise continuous then its discontinuities are restricted to a finite number of jump discontinuities in any finite interval of t, a jump discontinuity being one in which the change in $f(t)$ as we pass through the discontinuity is of finite magnitude. If $f(t)$ is piecewise monotonic then it is possible to divide any finite range of t into a finite number of intervals in such a way that within each interval $f(t)$ is either monotonically increasing or monotonically decreasing (i.e. $f(t)$ has no maxima or minima within the interval). We shall rarely have occasion to refer to these conditions in the present work.

The expression above for $f(t)$ represents the function $f(t)$ for all t. Evidently if we had a function $f(t)$ which was not periodic but which was defined only over the finite range of t, $-\pi \leqslant t \leqslant \pi$ then the expression would represent $f(t)$ for the values of t for which $f(t)$ is defined, i.e. $-\pi \leqslant t \leqslant \pi$.

More generally if the function $f(t)$ is periodic and has period T then the expansion becomes

$$f(t) = \tfrac{1}{2}a_0 + \sum_{n=1}^{\infty}\left(a_n \cos\frac{2\pi}{T}nt + b_n \sin\frac{2\pi}{T}nt\right)$$

A non-periodic function defined over a finite range $-\tfrac{1}{2}T \leqslant t \leqslant \tfrac{1}{2}T$ is also represented by this expression. Outside the range in which $f(t)$ is defined, i.e. $|t| > \tfrac{1}{2}T$ the expression represents a periodic extension of the function $f(t)$.

1.3 Properties of Sine and Cosine

We now list some important integral properties of sine and cosine functions which we shall need and which can readily be verified using elementary calculus. If p and q are integers (other than zero) then we have

$$\int_{-\pi}^{\pi} \cos pt\, dt = 0$$

$$\int_{-\pi}^{\pi} \sin pt\, dt = 0$$

$$\int_{-\pi}^{\pi} \cos^2 pt\, dt = \pi$$

$$\int_{-\pi}^{\pi} \sin^2 pt\, dt = \pi$$

$$\int_{-\pi}^{\pi} \cos pt . \cos qt \, dt = 0, \quad p \neq q$$

$$\int_{-\pi}^{\pi} \sin pt . \sin qt \, dt = 0, \quad p \neq q$$

$$\int_{-\pi}^{\pi} \cos pt . \sin qt \, dt = 0$$

The last three of these relations show that sines and cosines form an orthogonal set of functions. (They include the first two relations which are just special cases.)

1.4 Evaluation of the Coefficients

Using the relations given in the previous section we can evaluate the coefficients a_n, b_n. We have

$$f(t) = \tfrac{1}{2}a_0 + a_1 \cos t + a_2 \cos 2t + \ldots + b_1 \sin t + b_2 \sin 2t + \ldots$$

Hence integrating over the period $-\pi$ to π and using the relations of section 1.3 we have

$$\int_{-\pi}^{\pi} f(t) \, dt = \tfrac{1}{2}a_0 . 2\pi \qquad \therefore a_0 = \frac{1}{\pi}\int_{-\pi}^{\pi} f(t) \, dt$$

Similarly multiplying through by $\cos nt$ and integrating we have

$$\int_{-\pi}^{\pi} f(t) \cos nt \, dt = a_n \pi \qquad \therefore a_n = \frac{1}{\pi}\int_{-\pi}^{\pi} f(t) \cos nt \, dt$$

Multiplying through by $\sin nt$ and integrating we have

$$\int_{-\pi}^{\pi} f(t) \sin nt \, dt = b_n \pi \qquad \therefore b_n = \frac{1}{\pi}\int_{-\pi}^{\pi} f(t) \sin nt \, dt$$

For the more general case

$$\int_{-T/2}^{T/2} f(t) \, dt = \int_{-T/2}^{T/2} \tfrac{1}{2}a_0 \, dt$$

$$\therefore a_0 = \frac{2}{T}\int_{-T/2}^{T/2} f(t) \, dt$$

$$\int_{-T/2}^{T/2} f(t)\cos\frac{2\pi}{T}nt\,dt = \int_{-T/2}^{T/2} a_n\cos^2\frac{2\pi}{T}nt\,dt = \tfrac{1}{2}a_nT$$

$$\therefore\ a_n = \frac{2}{T}\int_{-T/2}^{T/2} f(t)\cos\frac{2\pi}{T}nt\,dt$$

and

$$\int_{-T/2}^{T/2} f(t)\sin\frac{2\pi}{T}nt\,dt = \int_{-T/2}^{T/2} b_n\sin^2\frac{2\pi}{T}nt\,dt = \tfrac{1}{2}b_nT$$

$$\therefore\ b_n = \frac{2}{T}\int_{-T/2}^{T/2} f(t)\sin\frac{2\pi}{T}nt\,dt$$

These expressions can be written in a still more general form since there is no necessity for the interval of integration to be symmetrical about the origin. The only requirement is that the integral shall be taken over a complete period. In general the lower limit can have any arbitrary value t_0 and the upper limit will then be t_0+T. However in the majority of the cases which we shall be considering we shall use the form quoted above employing the symmetrical interval.

1.5 Even and Odd Functions

If $f(t)$ is an even or an odd function then the series expansion has a special form. If $f(t)$ is even then by definition $f(-t) = f(t)$ and the expression $f(t)\sin nt$ is odd. Hence the integrand for b_n is odd. Now if we integrate an odd function over an interval which is symmetrical about the origin the result is zero, for if $\phi(t)$ is odd then

$$\int_{-\tau}^{\tau} \phi(t)\,dt = \int_{-\tau}^{0} \phi(t)\,dt + \int_{0}^{\tau} \phi(t)\,dt = \int_{\tau}^{0} \phi(-t)\,d(-t) + \int_{0}^{\tau} \phi(t)\,dt$$

$$= \int_{0}^{\tau} \phi(-t)\,dt + \int_{0}^{\tau} \phi(t)\,dt$$

and since $\phi(t)$ is odd $\phi(-t) = -\phi(t)$

$$\therefore \int_{-\tau}^{\tau} \phi(t)\,dt = -\int_{0}^{\tau} \phi(t)\,dt + \int_{0}^{\tau} \phi(t)\,dt = 0$$

Hence if the function $f(t)$ is even b_n is zero for all n and we have simply a cosine series. Similarly if $f(t)$ is odd then a_n is zero and we get a sine series.

Any function $f(t)$ can be expressed in terms of an even component $f_e(t)$ and an odd component $f_o(t)$ where

$$f_e(t) = \tfrac{1}{2}[f(t)+f(-t)]$$

and

$$f_o(t) = \tfrac{1}{2}[f(t)-f(-t)]$$

Thus in general we get a series containing both sines and cosines as already seen (section 1.2).

1.6 Complex Form of the Series

The occurrence of the sum of terms of the form $\cos nt$ and $\sin nt$ suggests the possibility of deriving a complex form of the series using the relation

$$\cos\theta + j\sin\theta = \exp(j\theta)$$

We have

$$f(t) = \tfrac{1}{2}a_0 + \sum_{n=1}^{\infty} (a_n \cos nt + b_n \sin nt)$$

Consider the two terms $a_n \cos nt + b_n \sin nt$. We can write

$$\begin{aligned} a_n \cos nt + b_n \sin nt &= \sqrt{(a_n^2+b_n^2)}\left\{\frac{a_n}{\sqrt{(a_n^2+b_n^2)}}\cos nt \right. \\ &\qquad \left. + \frac{b_n}{\sqrt{(a_n^2+b_n^2)}}\sin nt\right\} \\ &= \sqrt{(a_n^2+b_n^2)}\cos(nt-\phi_n) \end{aligned}$$

where

$$\tan\phi_n = \frac{b_n}{a_n}$$

Hence

$$a_n \cos nt + b_n \sin nt = \tfrac{1}{2}\sqrt{(a_n^2+b_n^2)}[\exp\{j(nt-\phi_n)\} + \exp\{-j(nt-\phi_n)\}]$$

So far we have considered n to have only positive integral values. If we allow n formally to have negative values also then we have from the expressions in the last section

$$a_{-n} = a_n, \quad b_{-n} = -b_n, \quad \text{and} \quad \phi_{-n} = -\phi_n$$

so that

$$a_n \cos nt + b_n \sin nt = \tfrac{1}{2}\sqrt{(a_n^2+b_n^2)}[\exp j\{nt-\phi_n\} + \exp j\{(-n)t-\phi_{(-n)}\}]$$

Thus we can now write

$$f(t) = \tfrac{1}{2}a_0 + \sum_{n=1}^{\infty} (a_n \cos nt + b_n \sin nt)$$

$$= \tfrac{1}{2}a_0 + \sum_{n=1}^{\infty} \tfrac{1}{2}\sqrt{(a_n^2+b_n^2)}[\exp j\{nt-\phi_n\} + \exp j\{(-n)t-\phi_{(-n)}\}]$$

$$= \tfrac{1}{2}a_0 + \sum_{n=1}^{\infty} \tfrac{1}{2}\sqrt{(a_n^2+b_n^2)} \exp j\{nt-\phi_n\} + \sum_{n=1}^{\infty} \tfrac{1}{2}\sqrt{(a_n^2+b_n^2)} \exp j\{(-n)t-\phi_{(-n)}\}$$

$$= \tfrac{1}{2}a_0 + \sum_{n=1}^{\infty} \tfrac{1}{2}\sqrt{(a_n^2+b_n^2)} \exp j\{nt-\phi_n\} + \sum_{n=-1}^{-\infty} \tfrac{1}{2}\sqrt{(a_n^2+b_n^2)} \exp j\{nt-\phi_n\}$$

$$= \tfrac{1}{2}a_0 + \left\{\sum_{n=1}^{\infty} + \sum_{n=-1}^{-\infty}\right\}\{\tfrac{1}{2}\sqrt{(a_n^2+b_n^2)} \exp j(nt-\phi_n)\}$$

and noting that for $n = 0$

$$\tfrac{1}{2}\sqrt{(a_n^2+b_n^2)} \exp j(nt-\phi_n) = \tfrac{1}{2}a_0, \quad \text{since } b_0 = \phi_0 = 0$$

we can write

$$f(t) = \sum_{n=-\infty}^{\infty} \tfrac{1}{2}\sqrt{(a_n^2+b_n^2)} \exp j(nt-\phi_n)$$

$$= \sum_{n=-\infty}^{\infty} \tfrac{1}{2}\sqrt{(a_n^2+b_n^2)} \exp(-j\phi_n) \exp(jnt)$$

or

$$f(t) = \sum_{n=-\infty}^{\infty} C_n \exp(jnt)$$

where

$$C_n = \tfrac{1}{2}\sqrt{(a_n^2+b_n^2)}\exp(-j\phi_n)$$

$$= \tfrac{1}{2}\sqrt{(a_n^2+b_n^2)}\left[\frac{a_n}{\sqrt{(a_n^2+b_n^2)}} - j\frac{b_n}{\sqrt{(a_n^2+b_n^2)}}\right]$$

$$\therefore C_n = \tfrac{1}{2}(a_n - jb_n)$$

$$= \frac{1}{2\pi}\int_{-\pi}^{\pi} f(t)\{\cos nt - j\sin nt\}\,dt$$

$$= \frac{1}{2\pi}\int_{-\pi}^{\pi} f(t)\exp(-jnt)\,dt$$

Thus we finally obtain the relations

$$f(t) = \sum_{n=-\infty}^{\infty} C_n \exp(jnt)$$

$$C_n = \frac{1}{2\pi}\int_{-\pi}^{\pi} f(t)\exp(-jnt)\,dt$$

In this section we have allowed n to assume negative values in a purely formal fashion. A negative value of n in effect represents a negative frequency and we shall return later to the significance of this concept.

As before we can extend this to the case of period T giving

$$f(t) = \sum_{n=-\infty}^{\infty} C_n \exp\left(j\frac{2\pi}{T}nt\right)$$

where

$$C_n = \frac{1}{T}\int_{t_0}^{t_0+T} f(t)\exp\left(-j\frac{2\pi}{T}nt\right)dt$$

1.7 Convergence of the Series

In order to justify the Fourier series expansion we shall show that the infinite series

$$\tfrac{1}{2}a_0 + \sum_{n=1}^{\infty} a_n \cos nt + b_n \sin nt$$

does converge to the value $f(t)$. Consider the finite series

$$\tfrac{1}{2}a_0 + a_1 \cos t + b_1 \sin t + a_2 \cos 2t + b_2 \sin 2t + \ldots + a_n \cos nt + b_n \sin nt$$

Define

$$a_n = \frac{1}{\pi}\int_{-\pi}^{\pi} f(t) \cos nt \, dt$$

and

$$b_n = \frac{1}{\pi}\int_{-\pi}^{\pi} f(t) \sin nt \, dt$$

where $f(t)$ is defined in the interval $-\pi \leqslant t \leqslant \pi$ and in this interval satisfies Dirichlet's conditions. Outside this interval $f(t)$ is defined by the relation $f(t+2\pi) = f(t)$ so that the function is periodic and has period 2π.

In the expressions defining a_n and b_n, t is a dummy variable. Hence we can write

$$a_n \cos nt + b_n \sin nt = \left\{\frac{1}{\pi}\int_{-\pi}^{\pi} f(x) \cos nx \, dx\right\} \cos nt + \left\{\frac{1}{\pi}\int_{-\pi}^{\pi} f(x) \sin nx \, dx\right\} \sin nt$$

where $f(x)$ is now defined for $-\pi \leqslant x \leqslant \pi$ and $f(x+2\pi) = f(x)$. This expression can be written

$$\frac{1}{\pi}\int_{-\pi}^{\pi} f(x)\{\cos nx \cos nt + \sin nx \sin nt\} \, dx = \frac{1}{\pi}\int_{-\pi}^{\pi} f(x) \cos n(x-t) \, dx$$

If $S_n(t)$ denotes the sum of the finite series then

$$S_n(t) = \frac{1}{\pi}\int_{-\pi}^{\pi} f(x)\{\tfrac{1}{2} + \cos(x-t) + \cos 2(x-t) + \ldots + \cos n(x-t)\} \, dx$$

Let $x-t=z$
and let

$$T_n = \tfrac{1}{2}+\cos z+\cos 2z+\ldots+\cos nz$$

$$\therefore\ 2\sin\tfrac{1}{2}z.T_n = \sin\tfrac{1}{2}z+2\sin\tfrac{1}{2}z.\cos z+\ldots+2\sin\tfrac{1}{2}z\cos nz$$

$$= \sin\tfrac{1}{2}z-\sin\tfrac{1}{2}z+\sin\tfrac{3}{2}z-\ldots-\sin(n-\tfrac{1}{2})z+\sin(n+\tfrac{1}{2})z$$

$$= \sin(n+\tfrac{1}{2})z$$

$$\therefore\ T_n = \frac{\sin(n+\tfrac{1}{2})z}{2\sin\tfrac{1}{2}z}$$

Hence

$$S_n(t) = \frac{1}{\pi}\int_{-\pi}^{\pi} f(x)\frac{\sin(n+\tfrac{1}{2})(x-t)}{2\sin\tfrac{1}{2}(x-t)}dx$$

We now investigate the behaviour of S_n as $n\to\infty$. Consider the integral

$$\int_a^b \phi(\zeta)\sin\alpha\zeta\,d\zeta$$

Suppose $\phi(\zeta)$ is finite and satisfies Dirichlet's conditions in the interval $a\leqslant\zeta\leqslant b$. When α becomes very large $\sin\alpha\zeta$ will go through a complete cycle with only a very small change in ζ and hence in $\phi(\zeta)$. If α becomes indefinitely large then $\phi(\zeta)$ can be considered constant whilst $\sin\alpha\zeta$ goes through one complete cycle and the integral will be zero, that is

$$\lim_{\alpha\to\infty}\int_a^b \phi(\zeta)\sin\alpha\zeta\,d\zeta = 0$$

Returning to the expression for $S_n(t)$ and putting $\tfrac{1}{2}(x-t)=\zeta$ we have

$$S_n(t) = \frac{1}{\pi}\int_{-\pi}^{\pi} f(x)\frac{\sin(n+\tfrac{1}{2})(x-t)}{2\sin\tfrac{1}{2}(x-t)}dx$$

$$= \frac{1}{\pi}\int_{-\frac{1}{2}(\pi+t)}^{\frac{1}{2}(\pi-t)} f(t+2\zeta)\frac{\sin(2n+1)\zeta}{2\sin\zeta}.2\,d\zeta$$

Comparing this with the expression above $\dfrac{f(t+2\zeta)}{\sin\zeta}$ corresponds to $\phi(\zeta)$. The only contribution to the integral (for large n) will be where the condition $\phi(\zeta)$ finite is violated, i.e. from the neighbourhood of points where $\sin\zeta = 0$. The range of integration is from $-\frac{1}{2}(\pi+t)$ to $\frac{1}{2}(\pi-t)$ and is thus dependent upon the value of the parameter t. We wish to consider values of t in the interval $-\pi \leqslant t \leqslant \pi$. Therefore the total range over which ζ will vary is also $-\pi$ to $+\pi$ and this includes three points where $\sin\zeta = 0$, viz. $\zeta = -\pi, 0, +\pi$. However we can exclude two of these points by considering the range $-\pi < t < \pi$ which does not include the two points $t = \pm\pi$. Hence for this range there is only one contribution to the integral namely that from the neighbourhood of the point $\zeta = 0$. We shall deal with this case first and then complete the proof by considering the two points at the ends of the range of t, $t = \pm\pi$ separately.

Since for the case $-\pi < t < \pi$ the only contribution to the integral is from the neighbourhood of the point $\zeta = 0$ we can take the range of integration as $-\delta$ to $+\delta$ so that

$$S_n(t) = \frac{1}{\pi}\int_{-\delta}^{\delta} f(t+2\zeta)\frac{\sin(2n+1)\zeta}{\sin\zeta}\,d\zeta$$

Furthermore in this interval we can put $\sin\zeta = \zeta$

$$\therefore\ S_n(t) = \frac{1}{\pi}\int_{-\delta}^{0} f(t+2\zeta)\frac{\sin(2n+1)\zeta}{\zeta}\,d\zeta + \frac{1}{\pi}\int_{0}^{\delta} f(t+2\zeta)\frac{\sin(2n+1)\zeta}{\zeta}\,d\zeta$$

Now as $\zeta \to 0$ from negative values $f(t+2\zeta)$ becomes $f(t_-)$, the limit of the function $f(t)$ at the point t when approached from the negative side. Putting $(2n+1)\zeta = \xi$ and letting $\zeta \to 0$

$$\frac{1}{\pi}\int_{-\delta}^{0} f(t+2\zeta)\frac{\sin(2n+1)\zeta}{\zeta}\,d\zeta = \frac{1}{\pi}f(t_-)\int_{-(2n+1)\delta}^{0}\frac{\sin\xi}{\xi}\,d\xi$$

As $n \to \infty$ this becomes

$$\frac{1}{\pi}f(t_-)\int_{-\infty}^{0}\frac{\sin\xi}{\xi}\,d\xi = \frac{1}{\pi}f(t_-)\frac{\pi}{2} = \tfrac{1}{2}f(t_-)$$

Similarly it may be shown that

$$\frac{1}{\pi}\int_0^{\delta} f(t+2\zeta)\frac{\sin(2n+1)\zeta}{\zeta}d\zeta = \tfrac{1}{2}f(t_+)$$

Hence we have finally

$$\lim_{n\to\infty} S_n(t) = \tfrac{1}{2}\{f(t_+)+f(t_-)\}$$

If the function $f(t)$ is continuous then $f(t_+) = f(t_-)$ and $S_n(t)\to f(t)$. At points of discontinuity S_n tends to the average value.

The expansion has been justified for $-\pi < t < \pi$. To complete the proof we must investigate the points $t = \pm\pi$. We require the values $\lim_{n\to\infty} S_n(\pm\pi)$. Consider first the value for $t = +\pi$. We have

$$S_n(t) = \frac{1}{\pi}\int_{-\pi}^{\pi} f(x)\frac{\sin(n+\frac{1}{2})(x-t)}{2\sin\frac{1}{2}(x-t)}dx$$

Substituting $\frac{1}{2}(x-t) = -\zeta$

$$S_n(t) = \frac{1}{\pi}\int_{\frac{1}{2}(t+\pi)}^{\frac{1}{2}(t-\pi)} f(t-2\zeta)\frac{\sin(2n+1)\zeta}{\sin\zeta}(-d\zeta)$$

$$= \frac{1}{\pi}\int_{\frac{1}{2}(t-\pi)}^{\frac{1}{2}(t+\pi)} f(t-2\zeta)\frac{\sin(2n+1)\zeta}{\sin\zeta}d\zeta$$

$$\therefore\ S_n(\pi) = \frac{1}{\pi}\int_0^{\pi} f(\pi-2\zeta)\frac{\sin(2n+1)\zeta}{\sin\zeta}d\zeta$$

Following the same procedure as before the only contribution to the integral will be where $\sin\zeta = 0$, i.e. in the neighbourhood of $\zeta = 0$ and $\zeta = \pi$.

$$\therefore\ S_n(\pi) = \frac{1}{\pi}\int_0^{\delta} f(\pi-2\zeta)\frac{\sin(2n+1)\zeta}{\sin\zeta}d\zeta$$

$$+\frac{1}{\pi}\int_{\pi-\delta'}^{\pi} f(\pi-2\zeta)\frac{\sin(2n+1)\zeta}{\sin\zeta}d\zeta$$

Substituting $\zeta' = \pi + \zeta$ in the second integral

$$S_n(\pi) = \frac{1}{\pi}\int_0^{\delta} f(\pi - 2\zeta)\frac{\sin(2n+1)\zeta}{\sin\zeta}\,d\zeta + \frac{1}{\pi}\int_{-\delta'}^{0} f(-\pi - 2\zeta')\frac{\sin(2n+1)\zeta'}{\sin\zeta'}\,d\zeta'$$

As $\zeta \to 0$ through positive values $f(\pi - 2\zeta)$ becomes $f(\pi_-)$ and $\sin\zeta \to \zeta$. As $\zeta' \to 0$ through negative values $f(-\pi - 2\zeta')$ becomes $f(-\pi_+)$ and $\sin\zeta' \to \zeta'$

$$\therefore\ S_n(\pi) = \frac{1}{\pi}f(\pi_-)\int_0^{\delta}\frac{\sin(2n+1)\zeta}{\zeta}\,d\zeta + \frac{1}{\pi}f(-\pi_+)\int_{-\delta'}^{0}\frac{\sin(2n+1)\zeta'}{\zeta'}\,d\zeta'$$

Finally putting $(2n+1)\zeta = \xi$ and $(2n+1)\zeta' = \xi'$ and letting $n \to \infty$ we obtain

$$S_n(\pi) = \frac{1}{\pi}f(\pi_-)\int_0^{\infty}\frac{\sin\xi}{\xi}\,d\xi + \frac{1}{\pi}f(-\pi_+)\int_{-\infty}^{0}\frac{\sin\xi'}{\xi'}\,d\xi'$$

$$= \frac{1}{\pi}f(\pi_-)\frac{\pi}{2} + \frac{1}{\pi}f(-\pi_+)\frac{\pi}{2}$$

$$\therefore\ S_n(\pi) = \tfrac{1}{2}\{f(\pi_-) + f(-\pi_+)\}$$

But since $f(t+2\pi) = f(t)$, $f(-\pi_+) = f(\pi_+)$

$$\therefore\ S_n(\pi) = \tfrac{1}{2}\{f(\pi_+) + f(\pi_-)\}$$

Similarly it can be shown that

$$S_n(-\pi) = \tfrac{1}{2}\{f(-\pi_+) + f(-\pi_-)\}$$

1.8 Multiplication Theorem

Using the complex form of the series we have

$$f(t) = \sum_{n=-\infty}^{\infty} C_n \exp(jnt)$$

$$C_n = \frac{1}{2\pi}\int_{-\pi}^{\pi} f(t)\exp(-jnt)\,dt$$

Consider the average value of the product $f_1(t).f_2(t)$ over a complete period. This is given by

$$\frac{1}{2\pi}\int_{-\pi}^{\pi} f_1(t)f_2(t)\,dt = \frac{1}{2\pi}\int_{-\pi}^{\pi}\left\{\sum_{n=-\infty}^{\infty}(C_1)_n \exp(jnt)\right\} f_2(t)\,dt$$

$$= \sum_{n=-\infty}^{\infty}(C_1)_n\left\{\frac{1}{2\pi}\int_{-\pi}^{\pi} f_2(t)\exp(jnt)\,dt\right\}$$

$$= \sum_{n=-\infty}^{\infty}(C_1)_n(C_2)_{-n}$$

i.e.

$$\frac{1}{2\pi}\int_{-\pi}^{\pi} f_1(t)f_2(t)\,dt = \sum_{n=-\infty}^{\infty}(C_1)_n(C_2)_{-n}$$

Thus the average value of the product of two functions over a complete period is equal to the sum of the products of the coefficients $(C_1)_n(C_2)_{-n}$ where the C's are the Fourier coefficients in the complex series representation of the functions.

1.9 Parseval's Theorem

From the multiplication theorem deduced in the previous section we have by putting $f_1 = f_2 = f$

$$\frac{1}{2\pi}\int_{-\pi}^{\pi}\{f(t)\}^2\,dt = \sum_{n=-\infty}^{\infty} C_n C_{-n}$$

Now from section 1.6 we have

$$C_n = \tfrac{1}{2}(a_n - jb_n)$$

Also we have $a_{-n} = a_n$ and $b_{-n} = -b_n$
Hence

$$C_{-n} = C_n^*$$

where C^* denotes the complex conjugate.

$$\therefore\ C_n C_{-n} = C_n C_n^* = |C_n|^2$$

so that

$$\frac{1}{2\pi}\int_{-\pi}^{\pi}\{f(t)\}^2\,dt = \sum_{n=-\infty}^{\infty}|C_n|^2$$

Writing C_n in terms of the components a_n, b_n

$$|C_n|^2 = \tfrac{1}{4}(a_n^2+b_n^2)$$

$$\therefore \sum_{n=-\infty}^{\infty}|C_n|^2 = \sum_{n=-\infty}^{\infty}\tfrac{1}{4}(a_n^2+b_n^2) = \tfrac{1}{4}a_0^2+\sum_{n=1}^{\infty}\tfrac{1}{2}(a_n^2+b_n^2)$$

$$\therefore \frac{1}{2\pi}\int_{-\pi}^{\pi}\{f(t)\}^2\,dt = \tfrac{1}{4}a_0^2+\sum_{n=1}^{\infty}\tfrac{1}{2}(a_n^2+b_n^2)$$

This is Parseval's theorem.

1.10 Power in Complex Waveform

Consider the term $a_n \cos nt$. The power in this waveform is given by

$$\frac{1}{2\pi}\int_{-\pi}^{\pi}(a_n\cos nt)^2\,dt$$

For example if a_n represents a voltage then the expression gives the power which would be developed in a resistance of unit value.

Now

$$\frac{1}{2\pi}\int_{-\pi}^{\pi}(a_n\cos nt)^2\,dt = \tfrac{1}{2}a_n^2$$

Similarly the power in the component $b_n \sin nt$ is given by $\frac{1}{2}b_n^2$. The power in the complex waveform $f(t)$ is given by

$$\frac{1}{2\pi}\int_{-\pi}^{\pi}\{f(t)\}^2\,dt$$

which by Parseval's theorem is equal to

$$\tfrac{1}{4}a_0^2+\sum_{n=1}^{\infty}\tfrac{1}{2}(a_n^2+b_n^2)$$

Hence the power in the complex waveform is equal to the sum of the powers in the Fourier components.

CHAPTER II

Analysis of Periodic Waveforms

2.1 Introduction

We have seen in Chapter I that a function $f(t)$ which is periodic, i.e. which is such that $f(t+T) = f(t)$ can be represented in terms of a Fourier series

$$f(t) = \tfrac{1}{2}a_0 + \sum_{n=1}^{\infty}\left(a_n \cos\frac{2\pi}{T}nt + b_n \sin\frac{2\pi}{T}nt\right)$$

In this chapter we shall apply these ideas to analyse some of the waveforms commonly occurring in the physical sciences.

2.2 The Square Wave

A sketch of this waveform is shown in Fig. 2.1. It is assumed that the amplitude of the wave is unity and the period T. With the choice of origin shown in Fig. 2.1 the function is anti-symmetric or odd.

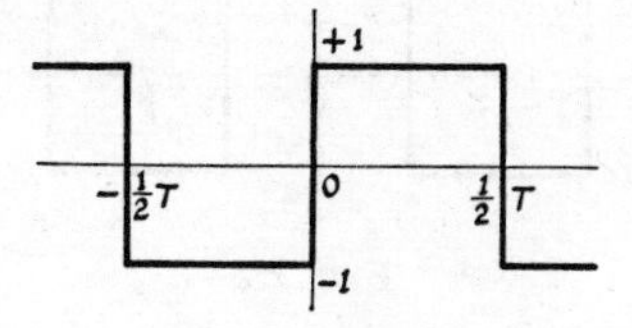

FIG. 2.1
The square wave—anti-symmetric form

Hence the Fourier expansion will be a sine series and we have $a_n = 0$ for all n (including $n = 0$). Using the expressions developed in Chapter I we have

$$b_n = \frac{2}{T}\int_{-T/2}^{T/2} f(t)\sin\frac{2\pi}{T}nt\,dt$$

Furthermore $f(t)$ is defined by the expressions

$$f(t) = -1, \quad -\tfrac{1}{2}T < t < 0$$
$$= 1, \quad 0 < t < \tfrac{1}{2}T$$

and
$$f(t+T) = f(t)$$

Therefore

$$b_n = \frac{2}{T}\left\{\int_{-T/2}^{0} (-1)\sin\frac{2\pi}{T}nt\,dt + \int_{0}^{T/2} (+1)\sin\frac{2\pi}{T}nt\,dt\right\}$$

$$= \frac{2}{T}\left\{\left[\frac{T}{2\pi n}\cos\frac{2\pi}{T}nt\right]_{-T/2}^{0} - \left[\frac{T}{2\pi n}\cos\frac{2\pi}{T}nt\right]_{0}^{T/2}\right\}$$

$$= \frac{1}{n\pi}\{[1-\cos(-n\pi)]+[1-\cos(n\pi)]\}$$

$$= \frac{2}{n\pi}(1-\cos n\pi)$$

$$\therefore b_n = 0, \quad n \text{ even}$$
$$= \frac{4}{n\pi}, \quad n \text{ odd}$$

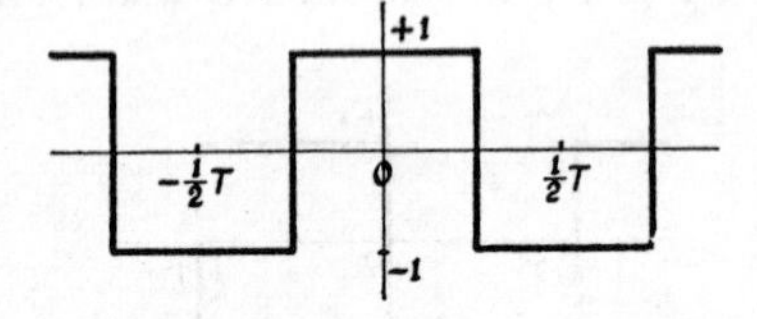

FIG. 2.2

The square wave—symmetric form

We could equally well have chosen the origin as shown in Fig. 2.2. In this case the function $f(t)$ would be defined as follows

$$f(t) = -1, \quad -\tfrac{1}{2}T < t < -\tfrac{1}{4}T$$
$$= +1, \quad -\tfrac{1}{4}T < t < \tfrac{1}{4}T$$
$$= -1, \quad \tfrac{1}{4}T < t < \tfrac{1}{2}T$$

and
$$f(t+T) = f(t)$$

The function is now even and hence $b_n = 0$. We have

$$a_0 = \frac{2}{T}\int_{-T/2}^{T/2} f(t)\,dt$$

$$= \frac{2}{T}\left\{\int_{-T/2}^{-T/4} (-1)\,dt + \int_{-T/4}^{T/4} (+1)\,dt + \int_{T/4}^{T/2} (-1)\,dt\right.$$

$$= 0$$

It can be seen that this must be the case since $\frac{1}{2}a_0$ represents the average value of $f(t)$ over one complete cycle (the d.c. term in electrical terminology) which is clearly zero. The rest of the coefficients are given by

$$a_n = \frac{2}{T}\int_{-T/2}^{T/2} f(t)\cos\frac{2\pi}{T}nt\,dt$$

$$= \frac{2}{T}\int_{-T/2}^{-T/4} (-1)\cos\frac{2\pi}{T}nt\,dt + \frac{2}{T}\int_{-T/4}^{T/4} (+1)\cos\frac{2\pi}{T}nt\,dt$$

$$+ \frac{2}{T}\int_{T/4}^{T/2} (-1)\cos\frac{2\pi}{T}nt\,dt$$

$$= \frac{1}{\pi n}\left\{-\left[\sin\frac{2\pi}{T}nt\right]_{-T/2}^{-T/4} + \left[\sin\frac{2\pi}{T}nt\right]_{-T/4}^{T/4} - \left[\sin\frac{2\pi}{T}nt\right]_{T/4}^{T/2}\right\}$$

$$= \frac{1}{\pi n}\left\{-\sin\left(-\frac{n\pi}{2}\right) + \sin\left(\frac{n\pi}{2}\right) - \sin\left(-\frac{n\pi}{2}\right) + \sin\left(\frac{n\pi}{2}\right)\right\}$$

$$= \frac{4}{\pi n}\sin\frac{n\pi}{2}$$

$$\therefore a_n = 0, \quad n \text{ even}$$

$$= +\frac{4}{n\pi}, \quad n = 1, 5, 9, \ldots \quad \text{or} \quad n = 4k-3$$

$$= -\frac{4}{n\pi}, \quad n = 3, 7, 11, \ldots \quad \text{or} \quad n = 4k-1$$

where $k = 1, 2, 3, \ldots$

By a suitable choice of origin we can expand the function either as a cosine series or as a sine series. The origin can of course be chosen anywhere yielding in general a series containing both sines and cosines.

2.3 Waveform Symmetry

In the previous section we deduced that for the square wave the even harmonics are all zero. It is interesting to investigate the condition

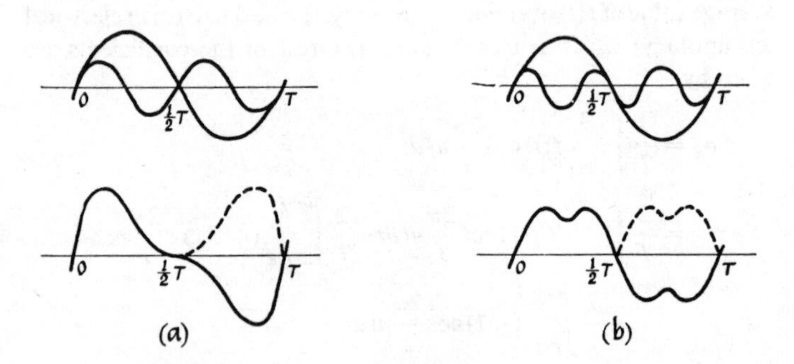

FIG. 2.3

The effects of Harmonic content (a) Odd harmonics; (b) even harmonics

for which this property occurs a little further. We can readily sketch the effect of the second harmonic and the third harmonic on the waveform of the fundamental. This is shown in Fig. 2.3. These two waveforms have a different type of symmetry. For the third harmonic case (Fig. 2.3b) the reflection of the second half of the waveform, (that between $\frac{1}{2}T$ and T), in the t axis produces a waveform (shown dotted) of exactly the same shape as that between 0 and $\frac{1}{2}T$. In the case of the second harmonic however the two corresponding waveforms are mirror images. We can express the type of symmetry exhibited in Fig. 2.3b by the relation

$$f(t+\tfrac{1}{2}T) = -f(t)$$

We then have

$$a_n = \frac{2}{T}\int_{-T/2}^{T/2} f(t)\cos\frac{2\pi}{T}nt\,dt$$

$$= \frac{2}{T}\int_{-T/2}^{0} f(t)\cos\frac{2\pi}{T}nt\,dt + \frac{2}{T}\int_{0}^{T/2} f(t)\cos\frac{2\pi}{T}nt\,dt$$

$$= \frac{2}{T}\int_{0}^{T/2} f(t+\tfrac{1}{2}T)\cos\frac{2\pi}{T}n(t+\tfrac{1}{2}T)\,dt + \frac{2}{T}\int_{0}^{T/2} f(t)\cos\frac{2\pi}{T}nt\,dt$$

$$= \frac{2}{T}\int_{0}^{T/2} -f(t)\cos\left(\frac{2\pi}{T}nt+n\pi\right)dt + \frac{2}{T}\int_{0}^{T/2} f(t)\cos\frac{2\pi}{T}nt\,dt$$

$$= \frac{2}{T}\int_{0}^{T/2} f(t)\left\{\cos\frac{2\pi}{T}nt-\cos\left(\frac{2\pi}{T}nt+n\pi\right)\right\}dt$$

$$= \frac{2}{T}\int_{0}^{T/2} f(t)\cos\frac{2\pi}{T}nt\,(1-\cos n\pi)\,dt$$

$$= 0, \quad n \text{ even}$$

Similarly

$$b_n = \frac{2}{T}\int_{-T/2}^{T/2} f(t)\sin\frac{2\pi}{T}nt\,dt$$

$$= \frac{2}{T}\int_{-T/2}^{0} f(t)\sin\frac{2\pi}{T}nt\,dt + \frac{2}{T}\int_{0}^{T/2} f(t)\sin\frac{2\pi}{T}nt\,dt$$

$$= \frac{2}{T}\int_{0}^{T/2} f(t+\tfrac{1}{2}T)\sin\frac{2\pi}{T}n(t+\tfrac{1}{2}T)\,dt + \frac{2}{T}\int_{0}^{T/2} f(t)\sin\frac{2\pi}{T}nt\,dt$$

$$= \frac{2}{T}\int_{0}^{T/2} -f(t)\sin\left(\frac{2\pi}{T}nt+n\pi\right)dt + \frac{2}{T}\int_{0}^{T/2} f(t)\sin\frac{2\pi}{T}nt\,dt$$

$$= \frac{2}{T}\int_{0}^{T/2} f(t)\left\{\sin\frac{2\pi}{T}nt-\sin\left(\frac{2\pi}{T}nt+n\pi\right)\right\}dt$$

$$= \frac{2}{T}\int_{0}^{T/2} f(t)\sin\frac{2\pi}{T}nt\,(1-\cos n\pi)\,dt$$

$$= 0, \quad n \text{ even}$$

Thus we see that waves exhibiting symmetry expressed by the relation $f(t+\frac{1}{2}T) = -f(t)$ have no even harmonic content. The push-pull amplifier is an example of a practical circuit which produces waveforms with no even harmonic distortion.

2.4 The Triangular Wave

A sketch of this waveform is shown in Fig. 2.4. This can be expressed analytically as

$$f(t) = 1+\frac{4}{T}t, \quad -\tfrac{1}{2}T < t < 0$$

$$= 1-\frac{4}{T}t, \quad 0 < t < \tfrac{1}{2}T$$

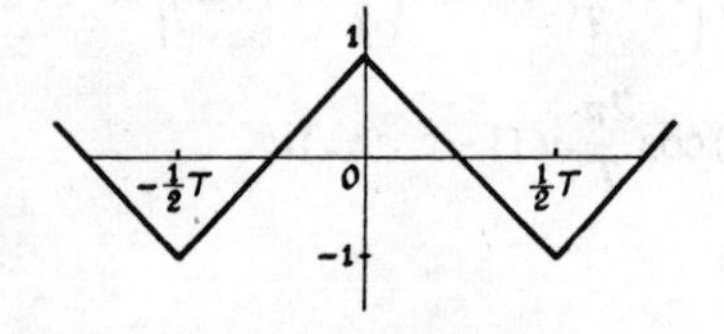

FIG. 2.4
The triangular wave

We see that $f(-t) = f(t)$, i.e. the function is even and hence $b_n = 0$.

$$a_0 = \frac{2}{T}\int_{-T/2}^{T/2} f(t)\,dt = \frac{2}{T}\int_{-T/2}^{0}\left(1+\frac{4}{T}t\right)dt+\frac{2}{T}\int_{0}^{T/2}\left(1-\frac{4}{T}t\right)dt = 0$$

This too could have been predicted since $\frac{1}{2}a_0$ represents the average value of $f(t)$ over one complete cycle (the d.c. term in electrical terminology) which is clearly zero.

$$a_n = \frac{2}{T}\int_{-T/2}^{T/2} f(t)\cos\frac{2\pi}{T}nt\,dt$$

$$= \frac{2}{T}\int_{-T/2}^{T/2}\cos\frac{2\pi}{T}nt\,dt+\frac{2}{T}\int_{-T/2}^{0}\frac{4}{T}t\cos\frac{2\pi}{T}nt\,dt$$

$$-\frac{2}{T}\int_{0}^{T/2}\frac{4}{T}t\cos\frac{2\pi}{T}nt\,dt$$

$$a_n = \frac{8}{T^2}\int_{+T/2}^{0} (-t)\cos\frac{2\pi}{T}nt\,d(-t) - \frac{8}{T^2}\int_{0}^{T/2} t\cos\frac{2\pi}{T}nt\,dt$$

$$= -\frac{16}{T^2}\int_{0}^{T/2} t\cos\frac{2\pi}{T}nt\,dt$$

$$= -\frac{16}{T^2}\left\{\left[t\frac{T}{2\pi n}\sin\frac{2\pi}{T}nt\right]_0^{T/2} - \frac{T}{2\pi n}\int_0^{T/2}\sin\frac{2\pi}{T}nt\,dt\right\}$$

$$= \frac{8}{T\pi n}\left[-\frac{T}{2\pi n}\cos\frac{2\pi}{T}nt\right]_0^{T/2}$$

$$= \frac{4}{\pi^2 n^2}(1-\cos n\pi)$$

$$\therefore a_n = \frac{8}{\pi^2 n^2}, \quad n \text{ odd}$$

$$= 0, \quad n \text{ even}$$

Note that this function satisfies the relation $f(t+\frac{1}{2}T) = -f(t)$ and hence all the even harmonics are zero as we showed generally in the last section.

2.5 The Saw-Tooth Wave

The saw-tooth is a waveform which occurs quite frequently in electronic engineering. It is sketched in Fig. 2.5, and can be represented by the function $f(t)$ defined by

$$f(t) = \frac{2}{T}t \qquad -\tfrac{1}{2}T < 0 < \tfrac{1}{2}T$$

$$f(t+T) = f(t)$$

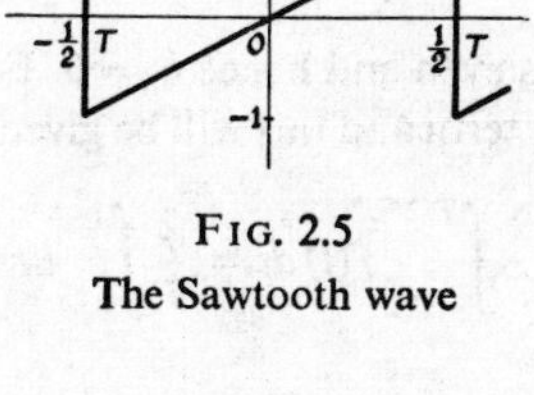

FIG. 2.5
The Sawtooth wave

In this case we have an odd function since $f(-t) = -f(t)$. Hence we have $a_n = 0$ and b_n given by

$$b_n = \frac{2}{T}\int_{-T/2}^{T/2} f(t)\sin\frac{2\pi}{T}nt\,dt = \frac{4}{T^2}\int_{-T/2}^{T/2} t\sin\frac{2\pi}{T}nt\,dt$$

$$= \frac{4}{T^2}\left\{\left[-\frac{T}{2\pi n}t\cos\frac{2\pi}{T}nt\right]_{-T/2}^{T/2} + \frac{T}{2\pi n}\int_{-T/2}^{T/2}\cos\frac{2\pi}{T}nt\,dt\right\}$$

$$= -\frac{2}{T\pi n}\left\{\frac{T}{2}\cos(n\pi) + \frac{T}{2}\cos(-n\pi)\right\}$$

$$= -\frac{2}{n\pi}\cos(n\pi)$$

$$\therefore b_n = +\frac{2}{n\pi}, \quad n \text{ odd}$$

$$= -\frac{2}{n\pi}, \quad n \text{ even}$$

2.6 The Infinite Train of Pulses

This is another commonly occurring waveform in practical technology. We shall first assume pulses of unit height, duration 2τ and period T.

The expression for $f(t)$ is

$$f(t) = 0 \qquad -\tfrac{1}{2}T < t < -\tau$$

$$= 1 \qquad -\tau < t < \tau$$

$$= 0 \qquad \tau < t < \tfrac{1}{2}T$$

$$f(t+T) = f(t)$$

The function $f(t)$ is even and hence $b_n = 0$. For the case shown in Fig. 2.6 there is a d.c. term and this will be given by $\frac{1}{2}a_0$ where

$$a_0 = \frac{2}{T}\int_{-T/2}^{T/2} f(t)\,dt = \frac{2}{T}\int_{-\tau}^{\tau} dt = \frac{4\tau}{T}$$

The rest of the coefficients a_n are given by

$$a_n = \frac{2}{T}\int_{-T/2}^{T/2} f(t)\cos\frac{2\pi}{T}nt\,dt = \frac{2}{T}\int_{-\tau}^{\tau} \cos\frac{2\pi}{T}nt\,dt$$

$$= \frac{1}{n\pi}\left[\sin\frac{2\pi}{T}nt\right]_{-\tau}^{\tau}$$

$$\therefore a_n = \frac{2}{n\pi}\sin\frac{2\pi}{T}n\tau$$

It is interesting to consider the limiting case of very narrow pulses, or impulses. Suppose we consider pulses of unit height and duration

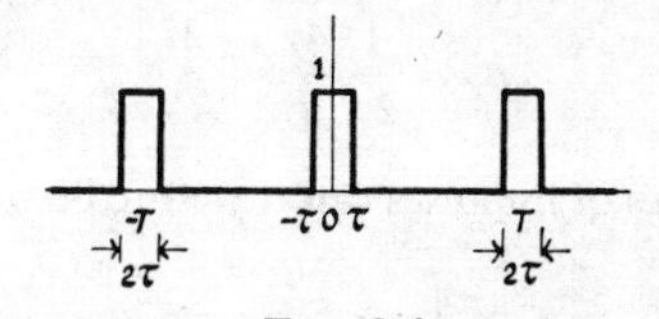

FIG. 2.6
A train of pulses

2τ and allow τ to become very small. In this case we can write

$$a_n = \frac{2}{n\pi}\sin\frac{2\pi}{T}n\tau \simeq \frac{2}{n\pi}\cdot\frac{2\pi}{T}n\tau = \frac{4\tau}{T}$$

and the magnitude of the a_n become independent of n. Thus the spectrum consists of an infinite series of lines of constant height and spacing. The original function of time we considered had exactly the same form. Functions of this type which have the same form when they are represented as a function of time as they have when represented as a function of frequency (the spectrum) are said to be self-reciprocal.

It may be noted that when $\tau \to 0$, $a_n \to 0$ and all the spectra have zero amplitude. This is because the original time waveform which is composed of spikes of finite amplitude is now represented as the sum of an infinite number of spectral terms all of which have become equally important. We can also look at this in terms of energy. If the original time waveform consists of spikes of finite amplitude and zero width then the energy in the waveform must be zero. We know furthermore from section 1.10 that this must be equal to the sum of the energies in the spectral components and hence these must all be zero.

We could form a train of *unit* impulses by defining the amplitude of the pulses as $1/2\tau$ instead of unity. The area under each pulse would then always be one. The spectral components will then have value

$$a_n = \frac{1}{2\tau}\frac{2}{n\pi}\sin\frac{2\pi}{T}n\tau$$

which as $\tau \to 0$ becomes

$$a_n = \frac{1}{2\tau}\cdot\frac{4\tau}{T} = \frac{2}{T}$$

The infinite series of impulses can therefore be written

$$f(t) = \frac{1}{T} + \frac{2}{T}\sum_{n=1}^{\infty}\cos\frac{2\pi}{T}nt$$

Under these conditions the amplitude of the original pulses becomes infinite, the energy per pulse remains finite, and since there are an infinite number of pulses in the train the total energy in the waveform is infinite also. In the frequency representation the amplitude of the individual components is finite, representing finite energy but again there is an infinite number of them so that the total energy is infinite.

2.7 The Rectifier

Rectifiers are used in electrical engineering to convert alternating currents and voltages into direct currents and voltages. The rectifier conducts only when the voltage is applied in one direction and not when the polarity is reversed. Thus if the applied voltage is sinusoidal the rectifier conducts only on the positive half cycles. It is common practice to use a combination of two or more rectifiers to produce what is known as full-wave rectification. In this scheme the negative half cycles are effectively reversed to produce a waveform as shown in Fig. 2.7.

Note that the period of the cosine wave from which the waveform of Fig. 2.7 is produced is $2T$. Thus it is seen that the fundamental component of the rectified cosine wave is double that of the original cosine wave. (In electrical terminology the 'hum' produced by full-wave rectifying the 50 c/s supply has a frequency of 100 c/s).

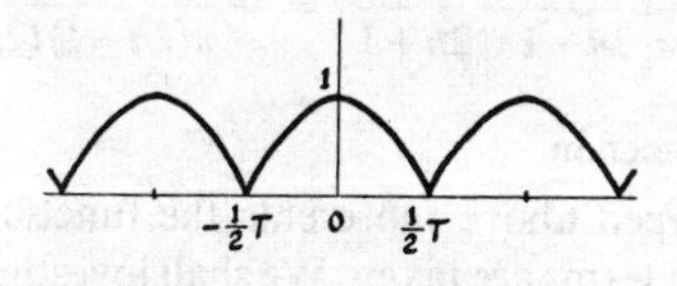

FIG. 2.7
A rectified cosine wave

The function representing this waveform is

$$f(t) = \cos\frac{\pi}{T}t, \quad -\tfrac{1}{2}T < t < \tfrac{1}{2}T$$

$$f(t+T) = f(t)$$

The d.c. component is given by $\frac{1}{2}a_0$ where

$$a_0 = \frac{2}{T}\int_{-T/2}^{T/2} \cos\frac{\pi}{T}t\,dt = \frac{2}{\pi}\left[\sin\frac{\pi}{T}t\right]_{-T/2}^{T/2}$$

$$\therefore a_0 = \frac{4}{\pi}$$

Thus the d.c. term is $2/\pi$.

The function is even and therefore $b_n = 0$. The harmonics are given by a_n where

$$a_n = \frac{2}{T}\int_{-T/2}^{T/2} \cos\left(\frac{\pi}{T}t\right)\cos\left(\frac{2\pi}{T}nt\right)dt$$

$$= \frac{1}{T}\int_{-T/2}^{T/2}\left\{\cos(2n-1)\frac{\pi}{T}t+\cos(2n+1)\frac{\pi}{T}t\right\}dt$$

$$= \frac{1}{\pi}\left[\frac{1}{2n-1}\sin(2n-1)\frac{\pi}{T}t+\frac{1}{2n+1}\sin(2n+1)\frac{\pi}{T}t\right]_{-T/2}^{T/2}$$

$$= \frac{2}{\pi}\left[\frac{1}{2n-1}\sin\left(n\pi-\frac{\pi}{2}\right)+\frac{1}{2n+1}\sin\left(n\pi+\frac{\pi}{2}\right)\right]$$

If n is even

$$a_n = \frac{2}{\pi}\left[\frac{-1}{2n-1}+\frac{1}{2n+1}\right] = -\frac{4}{\pi(2n-1)(2n+1)}$$

and if n is odd

$$a_n = \frac{2}{\pi}\left[\frac{1}{2n-1}-\frac{1}{2n+1}\right] = +\frac{4}{\pi(2n-1)(2n+1)}$$

2.8 Gibbs' Phenomenon

The series developed above represents the function $f(t)$ only if an infinite number of terms are taken. We shall investigate now the form of the curve when only a finite number of terms are taken. We can illustrate the phenomenon discovered by Gibbs by considering the square wave we investigated earlier (section 2.2).

For simplicity consider the square wave of unit amplitude and period 2π, i.e.

$$\begin{aligned} f(t) &= -1 \quad -\pi < t < 0 \\ &= +1 \quad 0 < t < \pi \end{aligned}$$

Consider the sum of a finite number of terms of the series. From section 1.7 this is given by

$$S_n(t) = \frac{1}{2\pi}\int_{-\pi}^{\pi} f(x)\frac{\sin(n+\frac{1}{2})(x-t)}{\sin\frac{1}{2}(x-t)}dx$$

Hence for the square wave we are considering

$$S_n(t) = \frac{1}{2\pi}\int_0^{\pi}\frac{\sin(n+\frac{1}{2})(x-t)}{\sin\frac{1}{2}(x-t)}dx - \frac{1}{2\pi}\int_{-\pi}^{0}\frac{\sin(n+\frac{1}{2})(x'-t)}{\sin\frac{1}{2}(x'-t)}dx'$$

Substituting $x-t=\theta$ and $x'-t=-\theta'$ we have

$$2\pi S_n(t) = \int_{-t}^{\pi-t}\frac{\sin(n+\frac{1}{2})\theta}{\sin\frac{1}{2}\theta}d\theta + \int_{\pi+t}^{t}\frac{\sin(n+\frac{1}{2})\theta'}{\sin\frac{1}{2}\theta'}d\theta'$$

Both θ and θ' are dummy variables in this expression. Using θ as the dummy variable throughout and rearranging we have

$$2\pi S_n(t) = \int_{-t}^{t}\frac{\sin(n+\frac{1}{2})\theta}{\sin\frac{1}{2}\theta}d\theta + \int_{\pi+t}^{\pi-t}\frac{\sin(n+\frac{1}{2})\theta}{\sin\frac{1}{2}\theta}d\theta$$

Now we know that when n becomes very large $S_n(t) \to f(t)$ except at the points of discontinuity, i.e. in this case at $t = 0$, when

$$S_n(t) \to \tfrac{1}{2}\{f(0_+)+f(0_-)\}$$

We shall now investigate how $S_n(t)$ behaves in the neighbourhood of the discontinuity, i.e. $t = 0$, when n is finite.

The first integral, for t small, is evaluated over the region where $\theta = 0$. At this point both the numerator and the denominator of the integrand become zero. The value of the integrand at $\theta = 0$ is obtained by taking the ratio of the derivatives of numerator and denominator, i.e.

$$[\{(n+\tfrac{1}{2})\cos(n+\tfrac{1}{2})\,\theta\}/\tfrac{1}{2}\cos\tfrac{1}{2}\theta]_{\theta=0} = 2n+1$$

The second integral is evaluated over the region where $\theta = \pi$. Here the magnitude of the integrand is approximately unity. Thus for all but the smallest values of n we can neglect the contribution from the second integral in comparison with that from the first. Thus for the conditions we are interested in investigating we can write

$$2\pi S_n(t) = \int_{-t}^{t} \frac{\sin(n+\tfrac{1}{2})\,\theta}{\sin\tfrac{1}{2}\theta}\,d\theta$$

Since the integrand is even in θ we have

$$2\pi S_n(t) = 2\int_{0}^{t} \frac{\sin(n+\tfrac{1}{2})\,\theta}{\sin\tfrac{1}{2}\theta}\,d\theta$$

and since we are interested in the region near the origin we can replace $\sin\tfrac{1}{2}\theta$ by $\tfrac{1}{2}\theta$ giving

$$2\pi S_n(t) = 2\int_{0}^{t} \frac{\sin(n+\tfrac{1}{2})\,\theta}{\tfrac{1}{2}\theta}\,d\theta$$

Substituting $n+\tfrac{1}{2} = m$

$$\pi S_n(t) = 2\int_{0}^{t} \frac{\sin m\theta}{\theta}\,d\theta$$

This is related to the sine integral, an important integral defined by the expression

$$Si(t) = \int_{0}^{t} \frac{\sin k}{k}\,dk$$

This integral is obviously zero when $t = 0$ and when $t \to \infty$ can be evaluated by the method of contour integration yielding a value $\frac{1}{2}\pi$. At other points it can be evaluated by numerical methods and has been tabulated. The form of the curve is shown in Fig. 2.8.

Considering now the form we deduced above

$$g(t) = \int_0^t \frac{\sin m\theta}{\theta} d\theta$$

Substitute $m\theta = k$

$$g(t) = \int_0^{mt} \frac{\sin k}{k} dk = Si(mt)$$

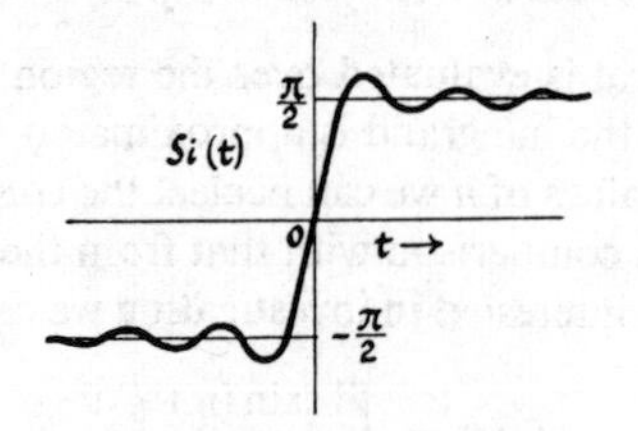

FIG. 2.8.

The Function $Si(t)$

Thus as m (or n) increases the curve is compressed into the origin but the general form of the curve does not change.

Returning now to the function we were considering, the finite series approximation to it is given by

$$S_n(t) = \frac{2}{\pi} \int_0^{(n+\frac{1}{2})t} \frac{\sin k}{k} dk$$

Thus $$S_n(0) = 0$$

and as $n \to \infty$ $$S_n(t) = \frac{2}{\pi} \int_0^{\infty} \frac{\sin k}{k} dk = 1$$

Hence at the origin the value of S_n is zero. It then rises rapidly as t increases and overshoots the value unity, the value of $f(t)$ in this region, passes through a maximum and oscillates about the line

$f(t) = 1$ with decreasing amplitude. The value of the overshoot is not dependent on n. As n increases the position at which the maximum occurs moves nearer to the point of discontinuity. If we consider a point removed from the discontinuity then as we have already mentioned as n increases the curve of Fig. 2.8 is compressed into the origin and therefore at the point under consideration the smaller the amplitude of the oscillation and the closer the value of S_n approaches one, the value of $f(t)$.

We can see more generally that S_n will produce an overshoot in the neighbourhood of a discontinuity as follows. Consider a function $f(t)$ which has a discontinuity at $t = 0$. Suppose for simplicity the function is odd. Let $f(0_+) = A$. We then have $f(0_-) = -A$ and $f(0) = 0$. Let $f(t)$ further be periodic with period 2π. The function $f(t)$ can be expressed as a sine series

$$f(t) = \sum_{r=1}^{\infty} b_r \sin rt$$

Hence

$$f(t) - S_n(t) = \sum_{r=n+1}^{\infty} b_r \sin rt$$

Now

$$\int_{-\pi}^{\pi} \{f(t) - S_n(t)\} \sin t\, dt = \sum_{r=n+1}^{\infty} \int_{-\pi}^{\pi} b_r \sin rt \sin t\, dt = 0$$

Since $f(t)$, $S_n(t)$ and $\sin t$ are all odd the integrand on the left hand side is even and hence

$$\int_{0}^{\pi} \{f(t) - S_n(t)\} \sin t\, dt = 0$$

Over the range of integration in this expression $\sin t$ is positive. Now the function $S_n(t)$ is the sum of a finite number of sinusoidal terms and hence $S_n(t)$ will approach zero when t is small. The function $f(t)$ on the other hand jumps discontinuously to a finite value. Hence for small positive values of t, $f(t) - S(t) < 0$ and will therefore give a negative contribution to the integral. This must be balanced by a positive contribution since the value of the integral is zero. Hence $S_n(t)$ must overshoot the value of $f(t)$ in some other region of the interval.

CHAPTER III

Fourier Integrals

3.1 Introduction

We saw in Chapter I that a function $f(t)$ defined over the region $-\frac{1}{2}T \leqslant t \leqslant \frac{1}{2}T$ could be represented as an infinite series

$$\tfrac{1}{2}a_0 + \sum_{n=1}^{\infty}\left(a_n \cos\frac{2\pi}{T}nt + b_n \sin\frac{2\pi}{T}nt\right)$$

or letting $2\pi/T = \omega$

$$\tfrac{1}{2}a_0 + \sum_{n=1}^{\infty}(a_n \cos n\omega t + b_n \sin n\omega t)$$

We can plot a_n and b_n as functions of frequency ω. These plots consist of a number of discrete spectral lines occurring at a constant spacing of $\omega = 2\pi/T$. As the region over which the function is defined becomes larger so the spacing of the lines becomes smaller. In the limit when the function is defined for all t, $T \to \infty$ and the spacing of the spectral lines becomes zero. Thus the plots of a_n and b_n are no longer discrete but become continuous and the sum becomes an integral. We shall now use this idea to derive the Fourier Integral Theorem as a limiting case of the series.

3.2 The Fourier Integral Theorem

If we consider a function $f(t)$ defined in the region $-\pi/\Omega \leqslant t \leqslant \pi/\Omega$ and obeying Dirichlet's conditions then we have already seen that we can represent $f(t)$ by the series

$$f(t) = \tfrac{1}{2}a_0 + \sum_{n=1}^{\infty}(a_n \cos n\Omega t + b_n \sin n\Omega t)$$

where

$$a_0 = \frac{\Omega}{\pi}\int_{-\pi/\Omega}^{\pi/\Omega} f(t)\, dt$$

$$a_n = \frac{\Omega}{\pi}\int_{-\pi/\Omega}^{\pi/\Omega} f(t)\cos n\Omega t\,dt$$

$$b_n = \frac{\Omega}{\pi}\int_{-\pi/\Omega}^{\pi/\Omega} f(t)\sin n\Omega t\,dt$$

Hence

$$f(t) = \frac{\Omega}{2\pi}\int_{-\pi/\Omega}^{\pi/\Omega} f(\lambda)\,d\lambda + \sum_{n=1}^{\infty}\frac{\Omega}{\pi}\left\{\cos n\Omega t\int_{-\pi/\Omega}^{\pi/\Omega} f(\lambda)\cos n\Omega\lambda\,d\lambda\right.$$

$$\left. + \sin n\Omega t\int_{-\pi/\Omega}^{\pi/\Omega} f(\lambda)\sin n\Omega\lambda\,d\lambda\right\}$$

$$= \frac{\Omega}{2\pi}\int_{-\pi/\Omega}^{\pi/\Omega} f(\lambda)\,d\lambda + \sum_{n=1}^{\infty}\frac{\Omega}{\pi}\int_{-\pi/\Omega}^{\pi/\Omega} f(\lambda)\cos n\Omega(t-\lambda)\,d\lambda$$

where λ is a dummy variable.

If we now take the limiting case $\Omega \to 0$ this expression becomes valid for all t. The spacing of the spectral lines in the plot against ω is Ω and hence as $\Omega \to 0$ the lines become more and more closely spaced. In the limit the spacing can be represented by $d\omega$ the increment of the continuous variable ω, and we get a continuous spectrum. Now n assumes all integral values from one to infinity and hence as $\Omega \to 0$, $n\Omega$ varies continuously from 0 to ∞ (provided we make $\Omega \to 0$ and $n \to \infty$ in such a way that $n\Omega \to \infty$). We can therefore represent $n\Omega$ by the continuous variable ω. The summation in the limit passes over to an integral and the limits as we have just seen will be 0 to ∞. Hence the second term in the expression can be written

$$\int_0^{\infty}\frac{d\omega}{\pi}\int_{-\infty}^{\infty} f(\lambda)\cos\omega(t-\lambda)\,d\lambda$$

The integral in the first term becomes $\int_{-\infty}^{\infty} f(\lambda)\,d\lambda$ in the limit and this is finite since $f(\lambda)$ satisfies Dirichlet's conditions. Hence as $\Omega \to 0$ the first term in the expression for $f(t)$ also tends to zero. Thus we have

$$f(t) = \frac{1}{\pi}\int_0^{\infty} d\omega\int_{-\infty}^{\infty} f(\lambda)\cos\omega(t-\lambda)\,d\lambda$$

This is Fourier's Integral Theorem.

3.3 Proof of the Integral Theorem

We shall now prove the Integral Theorem by a procedure analogous to that used in investigating the convergence of Fourier series. Consider the integral

$$I_A(t) = \int_0^A d\omega \int_{-\infty}^{\infty} f(\lambda) \cos \omega(t-\lambda)\, d\lambda$$

Reversing the order of integration we have

$$\begin{aligned} I_A(t) &= \int_{-\infty}^{\infty} f(\lambda)\, d\lambda \int_0^A \cos \omega(t-\lambda)\, d\omega \\ &= \int_{-\infty}^{\infty} f(\lambda)\, d\lambda \left[\frac{\sin \omega(t-\lambda)}{t-\lambda}\right]_0^A \\ &= \int_{-\infty}^{\infty} f(\lambda) \frac{\sin A(t-\lambda)}{t-\lambda}\, d\lambda \end{aligned}$$

We now let $A \to \infty$. Using again the theorem

$$\lim_{\alpha \to \infty} \int_a^b \phi(\zeta) \sin \alpha\zeta\, d\zeta = 0$$

provided $\phi(\zeta)$ is finite in the interval ab we see that the only contribution to the integral will be in the neighbourhood of the point where $t = \lambda$. Substituting $\lambda - t = \theta$ we have

$$I_A = \int_{-\infty}^{\infty} f(t+\theta) \frac{\sin(-A\theta)}{(-\theta)}\, d\theta = \int_{-\infty}^{\infty} f(t+\theta) \frac{\sin A\theta}{\theta}\, d\theta$$

As $A \to \infty$ the only contribution to this integral will be from the region near $\theta = 0$.

$$\begin{aligned} \therefore \lim_{A \to \infty} I_A &= \int_{-\delta}^{\delta} f(t+\theta) \frac{\sin A\theta}{\theta}\, d\theta \\ &= \int_{-\delta}^{0} f(t+\theta) \frac{\sin A\theta}{\theta}\, d\theta + \int_0^{\delta} f(t+\theta) \frac{\sin A\theta}{\theta}\, d\theta \\ &= \int_{-\delta}^{0} f(t_-) \frac{\sin A\theta}{\theta}\, d\theta + \int_0^{\delta} f(t_+) \frac{\sin A\theta}{\theta}\, d\theta \end{aligned}$$

Writing $A\theta = \phi$ we have

$$\lim_{A\to\infty} I_A = \int_{-\infty}^{0} f(t_-)\frac{\sin\phi}{\phi}\,d\phi + \int_0^{\infty} f(t_+)\frac{\sin\phi}{\phi}\,d\phi$$

$$= f(t_-)\int_{-\infty}^{0}\frac{\sin\phi}{\phi}\,d\phi + f(t_+)\int_0^{\infty}\frac{\sin\phi}{\phi}\,d\phi$$

$$= \frac{\pi}{2}\{f(t_+)+f(t_-)\}$$

Hence

$$\frac{1}{\pi}\int_0^{\infty} d\omega \int_{-\infty}^{\infty} f(\lambda)\cos\omega(t-\lambda)\,d\lambda = \tfrac{1}{2}\{f(t_+)+f(t_-)\}$$

If $f(t_+) = f(t_-)$, i.e. $f(t)$ is continuous at this point the expression becomes equal to $f(t)$ so that

$$\frac{1}{\pi}\int_0^{\infty} d\omega \int_{-\infty}^{\infty} f(\lambda)\cos\omega(t-\lambda)\,d\lambda = f(t)$$

3.4 Sine and Cosine Transforms

We have from the integral theorem

$$f(t) = \frac{1}{\pi}\int_0^{\infty} d\omega \int_{-\infty}^{\infty} f(\lambda)\cos\omega(t-\lambda)\,d\lambda$$

Expanding the cosine

$$f(t) = \frac{1}{\pi}\int_0^{\infty} d\omega \int_{-\infty}^{\infty} f(\lambda)\{\cos\omega t\cos\omega\lambda + \sin\omega t\sin\omega\lambda\}\,d\lambda$$

$$\therefore f(t) = \frac{1}{\pi}\int_0^{\infty}\cos\omega t\,d\omega\int_{-\infty}^{\infty} f(\lambda)\cos\omega\lambda\,d\lambda$$

$$+\frac{1}{\pi}\int_0^{\infty}\sin\omega t\,d\omega\int_{-\infty}^{\infty} f(\lambda)\sin\omega\lambda\,d\lambda$$

We define the Cosine Transform $a(\omega)$ as

$$a(\omega) = \int_{-\infty}^{\infty} f(\lambda)\cos\omega\lambda\,d\lambda$$

and the Sine Transform $b(\omega)$ as

$$b(\omega) = \int_{-\infty}^{\infty} f(\lambda) \sin \omega\lambda \, d\lambda$$

The function $f(t)$ can now be written in terms of these transforms.

$$f(t) = \frac{1}{\pi} \int_0^{\infty} \{a(\omega) \cos \omega t + b(\omega) \sin \omega t\} \, d\omega$$

The integrand contains the sum of a term in $\cos \omega t$ and $\sin \omega t$ so this can also be written in the amplitude-phase form

$$f(t) = \frac{1}{\pi} \int_0^{\infty} R(\omega) \cos \{\omega t + \phi(\omega)\} \, d\omega$$

where

$$R(\omega) = \sqrt{[\{a(\omega)\}^2 + \{b(\omega)\}^2]}$$

and

$$\tan \phi(\omega) = -\frac{b(\omega)}{a(\omega)}$$

Note that the region over which we integrate $a(\omega)$ and $b(\omega)$ is from 0 to ∞ and hence we are only concerned so far with the case $\omega > 0$, i.e. positive frequencies. In the next section we shall need to extend the range of ω to the negative values. The significance of negative frequencies will be discussed later.

3.5 Complex Transform

Using the results of the previous section we have

$$\begin{aligned} f(t) &= \frac{1}{\pi} \int_0^{\infty} \sqrt{[\{a(\omega)\}^2 + \{b(\omega)\}^2]} \cos \{\omega t + \phi(\omega)\} \, d\omega \\ &= \frac{1}{\pi} \int_0^{\infty} \sqrt{[\{a(\omega)\}^2 + \{b(\omega)\}^2]} \,.\, \tfrac{1}{2}[\exp\{j\omega t + j\phi(\omega)\} \\ &\qquad + \exp\{-j\omega t - j\phi(\omega)\}] \, d\omega \end{aligned}$$

We now formally allow ω to assume negative values and define $a(\omega)$ and $b(\omega)$ as in the previous section for all ω. We note that $a(\omega)$ is even and $b(\omega)$ and $\phi(\omega)$ are odd, i.e.

$$\begin{aligned} a(-\omega) &= a(\omega) \\ b(-\omega) &= -b(\omega) \\ \phi(-\omega) &= -\phi(\omega) \end{aligned}$$

Hence we can write

$$f(t) = \frac{1}{2\pi}\int_0^\infty \sqrt{(a^2+b^2)}[\exp(j\omega t)\exp\{j\phi(\omega)\}$$
$$+\exp(-j\omega t)\exp\{-j\phi(\omega)\}]\,d\omega$$
$$= \frac{1}{2\pi}\int_0^\infty \sqrt{(a^2+b^2)}[\exp(j\omega t)\exp\{j\phi(\omega)\}$$
$$+\exp\{j(-\omega)t\}\exp\{j\phi(-\omega)\}]\,d\omega$$
$$= \frac{1}{2\pi}\int_0^\infty \sqrt{(a^2+b^2)}\exp(j\omega t)\exp\{j\phi(\omega)\}\,d\omega$$
$$+\frac{1}{2\pi}\int_0^{-\infty} \sqrt{(a^2+b^2)}\exp(j\omega t)\exp\{j\phi(\omega)\}\,d(-\omega)$$

where ω has been replaced by $-\omega$ in the second integral. Hence

$$f(t) = \frac{1}{2\pi}\int_0^\infty \sqrt{(a^2+b^2)}\exp(j\omega t)\exp\{j\phi(\omega)\}\,d\omega$$
$$+\frac{1}{2\pi}\int_{-\infty}^0 \sqrt{(a^2+b^2)}\exp(j\omega t)\exp\{j\phi(\omega)\}\,d\omega$$
$$\therefore f(t) = \frac{1}{2\pi}\int_{-\infty}^\infty \sqrt{(a^2+b^2)}\exp\{j\phi(\omega)\}\exp(j\omega t)\,d\omega$$

Now define

$$F(\omega) = \sqrt{[\{a(\omega)\}^2+\{b(\omega)\}^2]}\exp\{j\phi(\omega)\}$$

Then

$$f(t) = \frac{1}{2\pi}\int_{-\infty}^\infty F(\omega)\exp(j\omega t)\,d\omega$$

The expression for $F(\omega)$ can be simplified

$$F(\omega) = \sqrt{(a^2+b^2)}\exp(j\phi) = \sqrt{(a^2+b^2)}\{\cos\phi+j\sin\phi\}$$
$$= \sqrt{(a^2+b^2)}\left[\frac{a}{\sqrt{(a^2+b^2)}}-j\frac{b}{\sqrt{(a^2+b^2)}}\right]$$
$$\therefore F(\omega) = a(\omega)-jb(\omega)$$

$F(\omega)$ can also be expressed in terms of $f(t)$

$$F(\omega) = a(\omega) - jb(\omega)$$

$$= \int_{-\infty}^{\infty} f(t)\cos\omega t\,dt - j\int_{-\infty}^{\infty} f(t)\sin\omega t\,dt$$

$$\therefore F(\omega) = \int_{-\infty}^{\infty} f(t)\exp(-j\omega t)\,dt$$

The function $F(\omega)$ is known as the complex Fourier transform of $f(t)$ and $f(t)$ is the inverse transform of $F(\omega)$. The two relations are expressed by the formulae

$$f(t) = \frac{1}{2\pi}\int_{-\infty}^{\infty} F(\omega)\exp(j\omega t)\,d\omega$$

$$F(\omega) = \int_{-\infty}^{\infty} f(t)\exp(-j\omega t)\,dt$$

The two functions $f(t)$ and $F(\omega)$ are known as a Fourier pair.

The Fourier transforms are special cases of the general integral transform $\bar{f}(\omega)$ of $f(t)$ (see for example Tranter, 1951), defined by

$$\bar{f}(\omega) = \int_a^b f(t)K(\omega, t)\,dt$$

The function $K(\omega, t)$ is called the kernel of the transform. In the case of the cosine transform $K(\omega, t) = \cos\omega t$, and the sine transform $K(\omega, t) = \sin\omega t$. In both cases $a = 0$ and $b = \infty$, so that

$$\bar{f}(\omega) = \int_0^{\infty} f(t)\cos\omega t\,dt \quad \text{Cosine Transform}$$

$$\bar{f}(\omega) = \int_0^{\infty} f(t)\sin\omega t\,dt \quad \text{Sine Transform}$$

For the complex Fourier transform $K(\omega, t) = \exp(-j\omega t)$, $a = -\infty$ and $b = \infty$

$$\bar{f}(\omega) = \int_{-\infty}^{\infty} f(t)\exp(-j\omega t)\,dt \quad \text{Complex Transform}$$

In general in this book we shall be concerned with functions $f(t)$ which are real functions of time. In this case $F(\omega)$ will in general be a complex function of the variable ω (which is real) and can be written in the form $a(\omega)-jb(\omega)$, where a and b are both real functions, as we have shown above. However there are cases of interest in which $f(t)$ may be complex; for example in modulation theory where we use the function $\exp(j\omega t).h(t)$, where h is real.

Since $F(\omega)$ is in general complex, if we wish to plot $F(\omega)$ as a function of frequency we must plot two graphs, one representing the real part and one representing the imaginary part, i.e. we plot $a(\omega)$ and $b(\omega)$. Note however that $b(\omega)$ is not the imaginary part of $F(\omega)$ since $F(\omega) = a(\omega)-jb(\omega)$. Thus $-b(\omega)$ is the imaginary part of $F(\omega)$.

We shall now consider some of the properties of $F(\omega)$. If $f(t)$ is an even function then $f(t)\sin\omega t$ is odd in t. Hence

$$b(\omega) = \int_{-\infty}^{\infty} f(t)\sin\omega t\,dt = 0$$

Thus for the case of an even function the complex transform $F(\omega)$ degenerates to a pure real function. Similarly if $f(t)$ is odd then $a(\omega) = 0$ and so for an odd function the transform $F(\omega)$ becomes pure imaginary.

We have already noted that $a(\omega)$ is even and $b(\omega)$ is odd. Hence if $f(t)$ is a real function so that $a(\omega)$ and $b(\omega)$ are real then we have

$$F(-\omega) = a(-\omega)-jb(-\omega) = a(\omega)+jb(\omega) = F^*(\omega)$$

so that

$$F^*(\omega) = F(-\omega)$$

where $F^*(\omega)$ denotes the complex conjugate of $F(\omega)$.

If $f(t)$ is complex then let

$$f(t) = h(t)+jk(t)$$

where $h(t)$ and $k(t)$ are real.

$$\begin{aligned}
F(\omega) &= \int_{-\infty}^{\infty} f(t)\exp(-j\omega t)\,dt \\
&= \int_{-\infty}^{\infty} \{h(t)+jk(t)\}(\cos\omega t - j\sin\omega t)\,dt \\
&= \int_{-\infty}^{\infty} [\{h(t)\cos\omega t + k(t)\sin\omega t\} \\
&\qquad + j\{k(t)\cos\omega t - h(t)\sin\omega t\}]\,dt
\end{aligned}$$

$$F(-\omega) = \int_{-\infty}^{\infty} [\{h(t)\cos\omega t - k(t)\sin\omega t\}$$
$$+j\{h(t)\sin\omega t + k(t)\cos\omega t\}]\,dt$$

$$F^*(\omega) = \int_{-\infty}^{\infty} [\{h(t)\cos\omega t + k(t)\sin\omega t\}$$
$$+j\{h(t)\sin\omega t - k(t)\cos\omega t\}]\,dt$$

Hence if $F(-\omega) = F^*(\omega)$ then we have the two conditions

$$\int_{-\infty}^{\infty} k(t)\sin\omega t\,dt = -\int_{-\infty}^{\infty} k(t)\sin\omega t\,dt$$

$$\therefore \int_{-\infty}^{\infty} k(t)\sin\omega t\,dt = 0$$

$$\therefore k(-t) = k(t)$$

and

$$\int_{-\infty}^{\infty} k(t)\cos\omega t\,dt = -\int_{-\infty}^{\infty} k(t)\cos\omega t\,dt$$

$$\therefore \int_{-\infty}^{\infty} k(t)\cos\omega t\,dt = 0$$

$$\therefore k(-t) = -k(t)$$

Hence from these two conditions we have

$$k(-t) = k(t) = -k(t)$$

$$k(t) = 0$$

Thus if $F(-\omega) = F^*(\omega)$ then $k(t) = 0$ and $f(t)$ is a real function. We therefore have the result that $F(-\omega) = F^*(\omega)$ if and only if $f(t)$ is a real function of t.

3.6 The Concept of Negative Frequencies

In the previous section we formally allowed ω to take negative values in order to develop the formula for the complex transform. A similar procedure was also used in the derivation of the complex form of the Fourier series. We shall now examine the concept of negative frequency a little more closely.

Since in each case the need for negative frequencies arose when we developed the complex form of the formula we shall start by discussing the complex representation of a sinusoidal type of variation, viz. $\exp(j\omega t)$. This is a complex function and can be written in terms of its components thus

$$\exp(j\omega t) = \cos\omega t + j\sin\omega t$$

This is a well-known relation. We can plot this on an Argand diagram in which the real part of the function is plotted along the x axis and the imaginary part along the y axis. The form of the function $\exp(j\omega t)$ shows that the amplitude is constant and its phase $\phi(t)$ is equal to ωt, i.e. it increases linearly at a rate determined by ω. Thus the function represents a point which moves round a circle in the Argand diagram at a constant rate in the positive direction of rotation, i.e. by convention counter-clockwise. It is now fairly easy to attach an interpretation to a negative frequency. Thus the phase of the function $\exp(-j\omega t)$ is $\phi(t) = -\omega t$, i.e. it increases at a constant rate but in the negative direction of rotation (clockwise). We sometimes think of a quantity which is specified by an amplitude and a phase as a vector. Hence we have the idea of representing a sinusoidal variation in terms of a rotating vector. If we require a real co-sinusoidal variation then we add together two vectors rotating at the same rate but in opposite directions, i.e. a positive and a negative frequency. This is represented analytically by the well-known relation

$$\cos\omega t = \tfrac{1}{2}\{\exp(j\omega t) + \exp(-j\omega t)\}$$

The complex spectrum of $\cos\omega t$ is pure real (we saw above that this is true for all even functions) and consists of two spikes one at $+\omega$ and one in the negative range of frequencies at $-\omega$. The spectrum is a line spectrum because $\cos\omega t$ is periodic. It is easily seen that the magnitude of the spectral lines is π since

$$\begin{aligned} f(t) &= \frac{1}{2\pi}\int_{-\infty}^{\infty} F(\omega)\exp(j\omega t)\,d\omega \\ &= \frac{1}{2\pi}[\pi\exp(j\omega_0 t) + \pi\exp(-j\omega_0 t)] \\ &= \tfrac{1}{2}\{\exp(j\omega_0 t) + \exp(-j\omega_0 t)\} = \cos\omega_0 t \end{aligned}$$

Note that both lines have magnitude π which is in agreement with the general result $a(-\omega) = a(\omega)$ obtained above.

It is worth while considering the spectrum of $\sin\omega t$. We wish to produce a real variation (i.e. along the x axis) which is sinusoidal. Again we produce the variation by adding two oppositely rotating vectors and they must be combined in such a way that the value of x at $t = 0$ is zero and, at first, x must increase as t increases. The vectors $\exp(j\omega t)$ and $\exp(-j\omega t)$ both have value 1 at $t = 0$. To make them cancel at this instant we multiply one by j and the other by $-j$, so that one lies along the direction of positive y and one along the direction of negative y. In order to make x increase with t at first the vector lying along the positive direction of y must rotate in the negative direction. Thus we require the sum $-j\exp(j\omega t)+j\exp(-j\omega t)$. If we specify unit amplitude then we must take half this sum. We note that this can be written

$$\frac{1}{2j}\{\exp(j\omega t) - \exp(-j\omega t)\}$$

which is the well-known formula for $\sin\omega t$. Thus the spectrum in this case is pure imaginary and consists of two lines one of magnitude $-\pi$ at $+\omega$ and one of $+\pi$ at $-\omega$. Note that this is the imaginary part of $F(\omega)$ and that if we had plotted $b(\omega)$ the signs would have been reversed, i.e. a positive value for positive frequency and a negative value for negative frequency since $F(\omega) = a(\omega) - jb(\omega)$. Again we see that the spectrum being pure imaginary is in agreement with the general result deduced above for odd functions and also the magnitude of the two lines are equal and opposite, which is in agreement with the general result $b(-\omega) = -b(\omega)$.

We can combine the results for $\cos\omega t$ and $\sin\omega t$ to obtain the spectrum of the exponential function. Thus we can write symbolically

$$f(t) = \sin\omega t, \quad F(\omega) = a(\omega) - jb(\omega)$$

$$= 0 - j\begin{bmatrix}\text{lines } +\pi \text{ at } +\omega \\ -\pi \text{ at } -\omega\end{bmatrix}$$

$$f(t) = j\sin\omega t, \quad F(\omega) = ja(\omega) + b(\omega)$$

$$= j.0 + \begin{bmatrix}\text{lines } +\pi \text{ at } +\omega \\ -\pi \text{ at } -\omega\end{bmatrix}$$

$$f(t) = \cos \omega t, \quad F(\omega) = a(\omega) - jb(\omega)$$

$$= \begin{bmatrix} \text{lines} + \pi \text{ at} + \omega \\ + \pi \text{ at} - \omega \end{bmatrix} + j.0$$

The quantities in the square brackets in the last two relations are real and hence can be added (see section 3.8 on the superposition theorem). Thus

$$f(t) = \cos \omega t + j \sin \omega t, \quad F(\omega) = [\text{line} + 2\pi \text{at} + \omega] + j.0$$

$$= \exp(j\omega t)$$

Thus the complex spectrum of $\exp(j\omega_0 t)$ is pure real and consists of a single line of magnitude 2π at $+\omega_0$. This is readily verified for

$$f(t) = \frac{1}{2\pi} \int_{-\infty}^{\infty} F(\omega) \exp(j\omega t)\, dt = \frac{1}{2\pi} [2\pi \exp(j\omega_0 t)] = \exp(j\omega_0 t)$$

Similarly the spectrum of $\exp(-j\omega_0 t)$ is a single line of magnitude 2π at $\omega = -\omega_0$.

3.7 The Shift Theorems

We shall now deduce a relationship between the spectrum of a waveform which has been shifted in time and the spectrum of the original waveform. Let $F(\omega)$ be the Fourier transform of $f(t)$ so that

$$F(\omega) = \int_{-\infty}^{\infty} f(t) \exp(-j\omega t)\, dt$$

and let $F_{-T}(\omega)$ be the Fourier transform of $f(t-T)$, i.e. the original waveform delayed by T. Then

$$F_{-T}(\omega) = \int_{-\infty}^{\infty} f(t-T) \exp(-j\omega t)\, dt$$

Putting $t - T = t'$

$$F_{-T}(\omega) = \int_{-\infty}^{\infty} f(t') \exp\{-j\omega(t'+T)\}\, dt'$$

$$= \exp(-j\omega T) \int_{-\infty}^{\infty} f(t') \exp(-j\omega t')\, dt'$$

Hence t' being a dummy variable we have

$$F_{-T}(\omega) = \exp(-j\omega T)\, F(\omega)$$

We can deduce formally a second shift theorem by a similar procedure, as follows. Let $f(t)$ be the inverse transform of $F(\omega)$ so that

$$f(t) = \frac{1}{2\pi}\int_{-\infty}^{\infty} F(\omega)\exp(j\omega t)\,d\omega$$

and let $f_{-p}(t)$ be the inverse transform of $F(\omega-p)$ so that

$$f_{-p}(t) = \frac{1}{2\pi}\int_{-\infty}^{\infty} F(\omega-p)\exp(j\omega t)\,d\omega$$

Putting $\omega-p=\omega'$

$$f_{-p}(t) = \frac{1}{2\pi}\int_{-\infty}^{\infty} F(\omega')\exp\{j(\omega'+p)t\}\,d\omega'$$

$$= \exp(jpt)\frac{1}{2\pi}\int_{-\infty}^{\infty} F(\omega')\exp(j\omega' t)\,d\omega'$$

Hence

$$f_{-p}(t) = \exp(jpt)f(t)$$

Expressing these two theorems in words:

(i) If the spectrum of $f(t)$ is $F(\omega)$, then the spectrum of $f(t-T)$ is $\exp(-j\omega T)\,F(\omega)$.

(ii) If the signal represented by $F(\omega)$ is $f(t)$, then the signal represented by $F(\omega-p)$ is $\exp(jpt)\,f(t)$.

We may note at this stage that both of these theorems have been deduced in a formal mathematical fashion. The physical interpretation of the first shift theorem presents no difficulty since if $f(t)$ is a real function of time then as already noted $f(t-T)$ represents the same signal delayed in time by amount T. The second shift theorem is not quite so straightforward however. We saw in section 3.4 that for a real function $f(t)$ the complex Fourier transform has the property $F^*(\omega)=F(-\omega)$, i.e. the real part of $F(\omega)$ is symmetrical (about the origin $\omega=0$) and the imaginary part is anti-symmetrical. If now we substitute $\omega-p$ for ω then this relation will clearly no longer be true since in general the function F cannot have the same symmetry properties about $\omega=p$. Hence the signal represented by $F(\omega-p)$ will not be a real function of t. The truth of this is evident from the expression deduced for $f_{-p}(t)$ above. Although in general we shall not be concerned with complex functions of time we shall use the second shift theorem when we come to consider modulation theory.

3.8 The Transform of a Derivative

If we have a function $f(t)$ whose Fourier transform is $F(\omega)$ then we can readily obtain the transform of the derivative of $f(t)$. We have

$$f(t) = \frac{1}{2\pi}\int_{-\infty}^{\infty} F(\omega)\exp(j\omega t)\,d\omega$$

and if we let $g(t)$ be the derivative of $f(t)$ then

$$g(t) = \frac{d}{dt}f(t) = \frac{1}{2\pi}\frac{d}{dt}\int_{-\infty}^{\infty} F(\omega)\exp(j\omega t)\,d\omega$$

Reversing the order of differentiation and integration and carrying out the differentiation gives

$$g(t) = \frac{1}{2\pi}\int_{-\infty}^{\infty} \{j\omega F(\omega)\}\exp(j\omega t)\,d\omega$$

We therefore see that $g(t)$ and $\{j\omega F(\omega)\}$ form a Fourier pair. The Fourier transform of the derivative of $f(t)$ is accordingly $j\omega$ times the Fourier transform of $f(t)$. The result is readily extended to higher order derivatives.

3.9 The Superposition Theorem

In this section we shall deduce an expression for the spectrum of the sum of two waveforms. Suppose the spectrum of $f_1(t)$ is $F_1(\omega)$ and that of $f_2(t)$ is $F_2(\omega)$. Then

$$f_1(t) = \frac{1}{2\pi}\int_{-\infty}^{\infty} F_1(\omega)\exp(j\omega t)\,d\omega$$

and

$$f_2(t) = \frac{1}{2\pi}\int_{-\infty}^{\infty} F_2(\omega)\exp(j\omega t)\,d\omega$$

Therefore

$$f_1(t)+f_2(t) = \frac{1}{2\pi}\int_{-\infty}^{\infty} F_1(\omega)\exp(j\omega t)\,d\omega + \frac{1}{2\pi}\int_{-\infty}^{\infty} F_2(\omega)\exp(j\omega t)\,d\omega$$

$$\therefore f_1(t)+f_2(t) = \frac{1}{2\pi}\int_{-\infty}^{\infty} \{F_1(\omega)+F_2(\omega)\}\exp(j\omega t)\,d\omega$$

Hence the spectrum of $\{f_1(t)+f_2(t)\}$ is $\{F_1(\omega)+F_2(\omega)\}$ or the spectrum of the sum of two waveforms is equal to the sum of their individual spectra. This is the superposition theorem.

3.10 The Duality Theorems

Let us consider a function of time $f(t)$ and a function of frequency $g(\omega)$, and suppose $f(t)$ and $g(\omega)$ form a Fourier pair. Then we can write

$$f(t) = \frac{1}{2\pi}\int_{-\infty}^{\infty} g(\omega)\exp(j\omega t)\,d\omega$$

$$\therefore f(-t) = \frac{1}{2\pi}\int_{-\infty}^{\infty} g(\omega)\exp(-j\omega t)\,d\omega$$

Hence by interchanging ω and t we obtain

$$f(-\omega) = \frac{1}{2\pi}\int_{-\infty}^{\infty} g(t)\exp(-j\omega t)\,dt$$

or

$$2\pi f(-\omega) = \int_{-\infty}^{\infty} g(t)\exp(-j\omega t)\,dt$$

Comparing this with the general relation

$$F(\omega) = \int_{-\infty}^{\infty} f(t)\exp(-j\omega t)\,dt$$

we see that $2\pi f(-\omega)$ and $g(t)$ are also a Fourier pair.

Similarly we could proceed from the relation

$$g(\omega) = \int_{-\infty}^{\infty} f(t)\exp(-j\omega t)\,dt$$

$$\therefore g(-\omega) = \int_{-\infty}^{\infty} f(t)\exp(j\omega t)\,dt$$

and interchanging ω and t

$$g(-t) = \int_{-\infty}^{\infty} f(\omega) \exp(j\omega t)\, d\omega$$

so that $\frac{1}{2\pi} g(-t)$ and $f(\omega)$ also form a Fourier pair.

We can state the duality theorems in words as follows:

(i) If the spectrum of $f(t)$ is $g(\omega)$ then the spectrum of $g(t)$ is $2\pi f(-\omega)$.

(ii) If the signal of $g(\omega)$ is $f(t)$ then the signal of $f(\omega)$ is $\frac{1}{2\pi} g(-t)$.

3.11 The Multiplication Theorem

We can obtain a formula for Fourier integrals corresponding to that derived in section 1.8 for Fourier series. Consider the integral of the product of two functions $f_1(t), f_2(t)$.

$$\begin{aligned}
\int_{-\infty}^{\infty} f_1(t) f_2(t)\, dt &= \frac{1}{2\pi} \int_{-\infty}^{\infty} f_1(t)\, dt \int_{-\infty}^{\infty} F_2(\omega) \exp(j\omega t)\, d\omega \\
&= \frac{1}{2\pi} \int_{-\infty}^{\infty} F_2(\omega)\, d\omega \int_{-\infty}^{\infty} f_1(t) \exp(j\omega t)\, dt \\
&= \frac{1}{2\pi} \int_{-\infty}^{\infty} F_1(-\omega) F_2(\omega)\, d\omega \\
&= \frac{1}{2\pi} \int_{-\infty}^{\infty} F_1^*(\omega) F_2(\omega)\, d\omega
\end{aligned}$$

Thus

$$\int_{-\infty}^{\infty} f_1(t) f_2(t)\, dt = \frac{1}{2\pi} \int_{-\infty}^{\infty} F_1^*(\omega) F_2(\omega)\, d\omega$$

Similarly it may be shown that

$$\int_{-\infty}^{\infty} f_1(t) f_2(t)\, dt = \frac{1}{2\pi} \int_{-\infty}^{\infty} F_1(\omega) F_2^*(\omega)\, d\omega$$

Hence we have

$$\int_{-\infty}^{\infty} f_1(t) f_2(t)\, dt = \frac{1}{2\pi} \int_{-\infty}^{\infty} F_1^*(\omega)\, F_2(\omega)\, d\omega$$

$$= \frac{1}{2\pi} \int_{-\infty}^{\infty} F_1(\omega)\, F_2^*(\omega)\, d\omega$$

3.12 Parseval's Theorem

As in the case of Fourier series we derive Parseval's theorem by putting $f_1 = f_2$ in the multiplication theorem. Hence we have

$$\int_{-\infty}^{\infty} \{f(t)\}^2\, dt = \frac{1}{2\pi} \int_{-\infty}^{\infty} F^*(\omega)\, F(\omega)\, d\omega$$

or

$$\int_{-\infty}^{\infty} \{f(t)\}^2\, dt = \frac{1}{2\pi} \int_{-\infty}^{\infty} |F(\omega)|^2\, d\omega$$

3.13 Energy Spectrum and Autocorrelation Function

The total energy in a waveform is proportional to the integral of the square of its amplitude

$$\int_{-\infty}^{\infty} \{f(t)\}^2\, dt$$

which by Parseval's theorem is equal to

$$\frac{1}{2\pi} \int_{-\infty}^{\infty} |F(\omega)|^2\, d\omega$$

The expression $|F(\omega)|^2$ is referred to as the energy spectrum. Now by the multiplication theorem

$$\int_{-\infty}^{\infty} f_1(t) f_2(t)\, dt = \frac{1}{2\pi} \int_{-\infty}^{\infty} F_1^*(\omega)\, F_2(\omega)\, d\omega$$

Let $f_2(t) = f_1(t+\tau)$. Then by the first shift theorem the transform of $f_1(t+\tau)$ is given by $\exp(j\omega\tau)\, F_1(\omega)$ where $F_1(\omega)$ is the transform of $f_1(t)$.

$$\therefore \int_{-\infty}^{\infty} f_1(t) f_1(t+\tau)\, dt = \frac{1}{2\pi} \int_{-\infty}^{\infty} F_1^*(\omega) \exp(j\omega\tau)\, F_1(\omega)\, d\omega$$

$$= \frac{1}{2\pi} \int_{-\infty}^{\infty} |F_1(\omega)|^2 \exp(j\omega\tau)\, d\omega$$

We define $\phi_{11}(\tau)$ by the relation

$$\phi_{11}(\tau) = \int_{-\infty}^{\infty} f_1(t) f_1(t+\tau)\, dt$$

The function $\phi_{11}(\tau)$ is known as the autocorrelation function. Hence we have

$$\phi_{11}(\tau) = \frac{1}{2\pi}\int_{-\infty}^{\infty} |F_1(\omega)|^2 \exp(j\omega\tau)\, d\omega$$

Thus $\phi_{11}(\tau)$ and $|F_1(\omega)|^2$ form a Fourier pair, and the energy spectrum is given by the Fourier transform of the autocorrelation function.

The autocorrelation function is of great importance in the analysis of random waveforms and also in probability theory. However a discussion of applications in these subjects is beyond the scope of the present work.

3.14 The Convolution Integral

Consider the Fourier transform of the product of two functions $f_1(t)$ and $f_2(t)$. If the Fourier transforms of $f_1(t)$ and $f_2(t)$ are $F_1(\omega)$ and $F_2(\omega)$ then the transform of the product, $F(\omega)$ may be obtained as follows.

$$\begin{aligned}
F(\omega) &= \int_{-\infty}^{\infty} f_1(t) f_2(t) \exp(-j\omega t)\, dt \\
&= \frac{1}{2\pi}\int_{-\infty}^{\infty} f_2(t) \exp(-j\omega t)\, dt \int_{-\infty}^{\infty} F_1(p) \exp(jpt)\, dp \\
&= \frac{1}{2\pi}\int_{-\infty}^{\infty} F_1(p)\, dp \int_{-\infty}^{\infty} f_2(t) \exp\{-j(\omega-p)t\}\, dt \\
&= \frac{1}{2\pi}\int_{-\infty}^{\infty} F_1(p) F_2(\omega-p)\, dp
\end{aligned}$$

and so we have the result

$$\int_{-\infty}^{\infty} f_1(t) f_2(t) \exp(-j\omega t)\, dt = \frac{1}{2\pi}\int_{-\infty}^{\infty} F_1(p) F_2(\omega-p)\, dp$$

The integral on the right hand side is known as the convolution integral. Thus the spectrum of the product of two functions is the convolution of the individual spectra of the two functions. The convolution operation is sometimes written symbolically

$$F(\omega) = \frac{1}{2\pi} F_1(\omega)*F_2(\omega)$$

We can also derive the dual of this theorem by taking the inverse transform of the product of two spectra.

$$\begin{aligned}
f(\tau) &= \frac{1}{2\pi}\int_{-\infty}^{\infty} F_1(\omega)\, F_2(\omega) \exp(j\omega\tau)\, d\omega \\
&= \frac{1}{2\pi}\int_{-\infty}^{\infty} F_2(\omega) \exp(j\omega\tau)\, d\omega \int_{-\infty}^{\infty} f_1(t) \exp(-j\omega t)\, dt \\
&= \frac{1}{2\pi}\int_{-\infty}^{\infty} f_1(t)\, dt \int_{-\infty}^{\infty} F_2(\omega) \exp\{j\omega(\tau - t)\}\, d\omega \\
&= \int_{-\infty}^{\infty} f_1(t) f_2(\tau - t)\, dt
\end{aligned}$$

and so

$$\begin{aligned}
\frac{1}{2\pi}\int_{-\infty}^{\infty} F_1(\omega)\, F_2(\omega) \exp(j\omega\tau)\, d\omega &= \int_{-\infty}^{\infty} f_1(t) f_2(\tau - t)\, dt \\
&= f_1(t) * f_2(t)
\end{aligned}$$

The signal of the product of two spectra is thus the convolution of the individual signals of the two spectra.

REFERENCE

TRANTER, C. J. *Integral Transforms in Mathematical Physics*, Methuen (1951)

CHAPTER IV

Analysis of Transients

4.1 Introduction

Just as Fourier series can be used in the analysis of periodic waveforms as seen in Chapter II so Fourier integrals can be used in the analysis of transient waveforms and in this chapter we shall consider some examples.

4.2 The Rectangular Pulse

Consider the rectangular pulse shown in Fig. 4.1(a) defined by the relations

$$f(t) = \frac{1}{2T} \qquad -T < t < +T$$

$$= 0 \quad \text{elsewhere}$$

Note that this definition results in a pulse of unit area. We obtain the spectrum of the pulse by taking the Fourier transform

$$F(\omega) = \int_{-\infty}^{\infty} f(t)\exp(-j\omega t)\,dt = \frac{1}{2T}\int_{-T}^{T} \exp(-j\omega t)\,dt$$

$$= \frac{1}{2T}\frac{(-1)}{j\omega}[\exp(-j\omega t)]_{-T}^{T}$$

$$= \frac{1}{2T}\frac{1}{j\omega}\{\exp(j\omega T)-\exp(-j\omega T)\}$$

or $$F(\omega) = \frac{\sin \omega T}{\omega T}$$

This expression will be zero whenever $\sin \omega T = 0$, i.e. for $\omega = n\pi/T$ except at the origin when the denominator becomes zero also. When

ωT is small $\sin \omega T$ is approximately equal to ωT and the value of $F(\omega)$ at $\omega = 0$ becomes unity. As ω becomes larger $\sin \omega T$ continues to oscillate between $+1$ and -1 whereas ωT the denominator

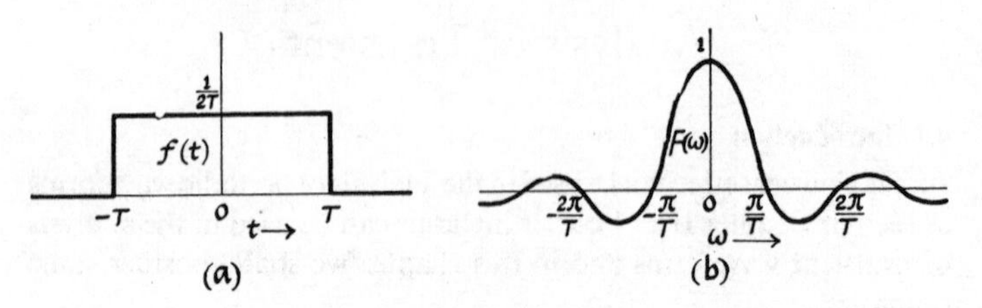

FIG. 4.1

(*a*) The rectangular pulse and (*b*) its spectrum

increases continuously. Hence as ω increases the magnitude of the oscillations decrease and finally become zero. A sketch of the spectrum is shown in Fig. 4.1(*b*).

4.3 The Rectangular Frequency Distribution

We shall consider now the analogous case of a rectangular frequency distribution (Fig. 4.2) defined by

$$F(\omega) = \frac{1}{2\omega_c} \qquad -\omega_c < \omega < +\omega_c$$

$$= 0 \qquad \text{elsewhere}$$

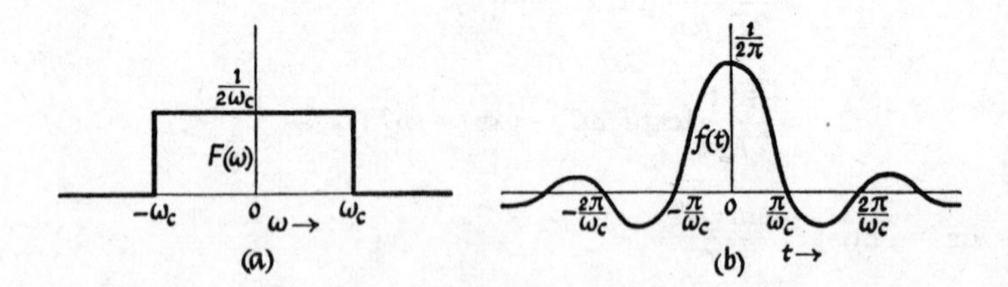

FIG. 4.2

(*a*) The rectangular frequency distribution and (*b*) its signal

We have

$$f(t) = \frac{1}{2\pi}\int_{-\infty}^{\infty} F(\omega)\exp(j\omega t)\,d\omega$$

$$= \frac{1}{2\pi}\frac{1}{2\omega_c}\int_{-\omega_c}^{\omega_c} \exp(j\omega t)\,d\omega$$

$$= \frac{1}{2\pi}\frac{1}{2\omega_c}\frac{1}{jt}[\exp(j\omega t)]_{-\omega_c}^{\omega_c}$$

$$= \frac{1}{4\pi j\omega_c t}\{\exp(j\omega_c t) - \exp(-j\omega_c t)\}$$

$$= \frac{1}{2\pi}\frac{\sin\omega_c t}{\omega_c t}$$

We could have obtained this result from the duality theorem and the results of the last section. The duality theorem states that if the signal of $g(\omega)$ is $f(t)$ then the signal of $f(\omega)$ is $g(-t)/2\pi$. Thus we define $g(\omega)$ by the relation

$$g(\omega) = \frac{\sin\omega T}{\omega T}$$

and

$$f(t) = \frac{1}{2T} \qquad -T < t < +T$$

$$= 0 \qquad \text{elsewhere}$$

Thus $f(t)$ is the signal of $g(\omega)$. The function $f(\omega)$ would then be defined by

$$f(\omega) = \frac{1}{2\omega_c} \qquad -\omega_c < \omega < \omega_c$$

$$= 0 \qquad \text{elsewhere}$$

where we have replaced the particular value of time, T, by the particular value of frequency ω_c. Hence by the duality theorem the signal of this spectrum is given by

$$\frac{1}{2\pi}g(-t) = \frac{1}{2\pi}\frac{\sin(-\omega_c t)}{(-\omega_c t)} = \frac{1}{2\pi}\frac{\sin\omega_c t}{\omega_c t}$$

where again T is replaced by ω_c.

4.4 Reciprocal Spreading

We have seen in the discussion above that the first zero in the frequency spectrum of the signal confined to the region $-T < t < T$ in time occurs at a frequency $\omega = \pi/T$. Since the majority of the signal energy will be contained in frequencies less than this we can consider the bandwidth of the signal to be of this order, i.e. we may write $\Delta f \sim 1/2T$. The time Δt occupied by the signal is $2T$ and hence we have

$$\Delta f . \Delta t \sim 1$$

Hence as the region in time occupied by the signal becomes smaller so the region it occupies in frequency becomes larger. A similar result is obtained by considering the signal confined to the region of frequencies $-\omega_c < \omega < \omega_c$, i.e. frequencies less than ω_c so that $\Delta f = \omega_c/2\pi$. Here the signal energy lies mainly in the time interval $-\pi/\omega_c < t < \pi/\omega_c$ so that $\Delta t = 2\pi/\omega_c$ again giving $\Delta f . \Delta t \sim 1$. A signal can be represented either as a function of time or as a function of frequency (i.e. its spectrum) and as it is compressed in one representation so it expands in the other. This phenomenon is sometimes referred to as reciprocal spreading.

4.5 The Unit Impulse

This is an important function and can be developed from the rectangular pulse of section 4.2. This was defined in such a way that its area was unity. The unit impulse is obtained by taking the limit $T \to 0$. The amplitude of the function then becomes infinite at the origin but the area remains unity. The spectrum retains the value one at the origin $\omega = 0$ and the first zero on each side of the origin move out to infinity. The spectrum then becomes

$$F(\omega) = \lim_{T \to 0} \frac{\sin \omega T}{\omega T} = 1$$

i.e. the amplitude of the spectrum is independent of ω and becomes a straight line parallel to the ω axis at amplitude one. The energy of the unit impulse is infinite.

By the duality theorem or from section 4.3 we deduce that a frequency impulse at the origin gives a signal of constant height $1/2\pi$, i.e. a frequency impulse at the origin represents a d.c. signal, which is obviously correct.

Now applying the second shift theorem a frequency impulse occurring at frequency $\omega = p$ will represent a signal $\exp(jpt)f(t)$. This, as was explained in the discussion on the shift theorem is not a real function of time. We can overcome this difficulty however as we saw in section 3.6 by adding a second impulse at frequency $\omega = -p$ which by the shift theorem gives a signal $\exp(-jpt)f(t)$. Since $f(t)$ was a constant $(1/2\pi)$ the resultant signal is $(1/\pi)\cos pt$, by the superposition theorem. Hence a sinusoidal type of variation gives rise to frequency impulses or spectral lines, a result we are already familiar with. Note that for the signal $\cos pt$ the magnitude of the spectral lines will be π, for

$$f(t) = \frac{1}{2\pi}\int_{-\infty}^{\infty} F(\omega)\exp(j\omega t)\,d\omega$$

$$= \frac{1}{2\pi}\{\pi\exp(jpt) + \pi\exp(-jpt)\} = \cos pt$$

as required (see also section 3.6).

4.6 Finite Train of Pulses

From section 4.2 we have that the spectrum of a pulse of unit area and duration 2τ at the origin is $\sin(\omega\tau)/(\omega\tau)$. Thus a pulse occurring

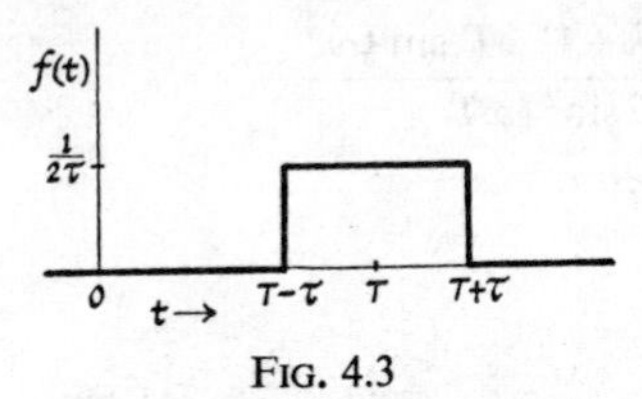

FIG. 4.3
A rectangular pulse occurring at time T

at time $t = T$ as shown in Fig. 4.3 will have a spectrum given by $\exp(j\omega T).\sin(\omega\tau)/(\omega\tau)$, by the shift theorem. Now consider a train of N pulses occurring at

$$t = -kT, -(k-1)T, \ldots -T, 0, T, \ldots (k-1)T, kT$$

where $N = 2k+1$. Then by the superposition theorem the spectrum is given by

$$F(\omega) = \frac{\sin \omega\tau}{\omega\tau}[1+\exp(j\omega T)+\exp(-j\omega T)+\ldots$$

$$+\exp(jk\omega T)+\exp(-jk\omega T)]$$

$$= \frac{\sin \omega\tau}{\omega\tau}[-1+2\{1+\cos\omega T+\cos 2\omega T+\ldots+\cos k\omega T\}]$$

The series in square brackets may be summed by taking the real part of the corresponding exponential series, i.e.

$$-1+2\,\mathrm{Re}[1+\exp(j\omega T)+\ldots+\exp(jk\omega T)]$$

$$= -1+2\,\mathrm{Re}\left[\frac{1-\exp\{j(k+1)\,\omega T\}}{1-\exp(j\omega T)}\right]$$

$$= -1+2\,\mathrm{Re}\left[\frac{\{1-\exp j(k+1)\,\omega T\}\{1-\exp(-j\omega T)\}}{2(1-\cos\omega T)}\right]$$

$$= -1+\frac{1-\cos(\omega T)+\cos(k\omega T)-\cos(k+1)\,\omega T}{1-\cos\omega T}$$

$$= \frac{\cos(k\omega T)-\cos(k+1)\,\omega T}{1-\cos\omega T}$$

$$= \frac{2\sin\frac{1}{2}(2k+1)\,\omega T.\sin\frac{1}{2}\omega T}{2\sin^2\frac{1}{2}\omega T}$$

$$= \frac{\sin\frac{1}{2}N\omega T}{\sin\frac{1}{2}\omega T}$$

Hence we have

$$F(\omega) = \frac{\sin\omega\tau}{\omega\tau}\frac{\sin\frac{1}{2}N\omega T}{\sin\frac{1}{2}\omega T}$$

which is the spectrum of the finite train of pulses.

Let us now consider the behaviour as $N\to\infty$. We can write

$$f(t) = \frac{1}{2\pi}\int_{-\infty}^{\infty}\frac{\sin\omega\tau}{\omega\tau}\frac{\sin\frac{1}{2}N\omega T}{\sin\frac{1}{2}\omega T}\exp(j\omega t)\,d\omega$$

Using again the theorem $\int_a^b \phi(\zeta)\sin\alpha\zeta\, d\zeta \to 0$ as $\alpha\to\infty$ provided $\phi(\zeta)$ is finite in the region $a<\zeta<b$ we see that the only contribution to the integral will be where the function

$$\phi(\omega) = \frac{\sin\omega\tau}{\omega\tau}\frac{1}{\sin\frac{1}{2}\omega T}\exp(j\omega t)$$

is not finite, i.e. where

$$\sin\tfrac{1}{2}\omega T = 0$$

This occurs when

$$\tfrac{1}{2}\omega T = n\pi \quad \text{or} \quad \omega = \frac{2\pi n}{T}$$

The contribution from one such point will be

$$f_n(t) = \frac{1}{2\pi}\int_{(2\pi n/T)-\delta}^{(2\pi n/T)+\delta}\frac{\sin(2\pi n\tau/T)}{(2\pi n\tau/T)}\exp\left(j\frac{2\pi nt}{T}\right)\frac{\sin\frac{1}{2}N\omega T}{\sin\frac{1}{2}\omega T}\,d\omega$$

where ω has been replaced by the value $2\pi n/T$ in those functions which are finite and slowly varying in the region of integration.

$$\therefore f_n(t) = \frac{1}{2\pi}\frac{\sin(2\pi n\tau/T)}{(2\pi n\tau/T)}\exp\left(j\frac{2\pi nt}{T}\right)\int_{(2\pi n/T)-\delta}^{(2\pi n/T)+\delta}\frac{\sin\frac{1}{2}N\omega T}{\sin\frac{1}{2}\omega T}\,d\omega$$

Now let $\frac{1}{2}N\omega T = \xi$. Foı large N we have

$$\int_{(2\pi n/T)-\delta}^{(2\pi n/T)+\delta}\frac{\sin\frac{1}{2}N\omega T}{\sin\frac{1}{2}\omega T}\,d\omega = \int_{-\infty}^{\infty}\frac{\sin\xi}{\sin(\xi/N)}\cdot\frac{2}{NT}\,d\xi = \int_{-\infty}^{\infty}\frac{\sin\xi}{(\xi/N)}\frac{2}{NT}\,d\xi$$

$$= \frac{2}{T}\int_{-\infty}^{\infty}\frac{\sin\xi}{\xi}\,d\xi = \frac{2\pi}{\cdot T}$$

$$\therefore f_n(t) = \frac{1}{2\pi}\frac{\sin(2\pi n\tau/T)}{(2\pi n\tau/T)}\exp\left(j\frac{2\pi nt}{T}\right)\frac{2\pi}{T}$$

$$= \frac{1}{2\pi}\frac{\sin(2\pi n\tau/T)}{n\tau}\exp\left(j\frac{2\pi nt}{T}\right)$$

Hence $f(t)$ is given by

$$f(t) = \sum_{n=-\infty}^{\infty} \frac{1}{2\pi} \frac{\sin(2\pi n\tau/T)}{n\tau} \exp\left(j\frac{2\pi nt}{T}\right)$$

$$= \frac{1}{2\pi}\left[\frac{2\pi}{T} + \sum_{n=1}^{\infty} \frac{\sin(2\pi n\tau/T)}{n\tau}\left\{\exp\left(j\frac{2\pi nt}{T}\right) + \exp\left(-j\frac{2\pi nt}{T}\right)\right\}\right]$$

$$\therefore f(t) = \frac{1}{T} + \frac{1}{\pi}\sum_{n=1}^{\infty} \frac{\sin(2\pi n\tau/T)}{n\tau} \cos\frac{2\pi}{T}nt$$

This result could of course have been obtained more easily by the Fourier series method

$$a_0 = \frac{2}{T}\int_{-\tau}^{\tau} \frac{1}{2\tau}\,dt = \frac{2}{T}$$

$$a_n = \frac{2}{T}\int_{-\tau}^{\tau} \frac{1}{2\tau} \cos\frac{2\pi}{T}nt\,dt = \frac{1}{\pi}\frac{\sin(2\pi n\tau/T)}{n\tau}$$

Hence as before

$$f(t) = \frac{1}{T} + \frac{1}{\pi}\sum_{n=1}^{\infty} \frac{\sin(2\pi n\tau/T)}{n\tau} \cos\frac{2\pi}{T}nt$$

4.7 The Gaussian Function

Consider the simple form of the Gaussian function

$$f(t) = \exp(-\tfrac{1}{2}t^2)$$

The Fourier transform is given by

$$F(\omega) = \int_{-\infty}^{\infty} \exp(-\tfrac{1}{2}t^2)\exp(-j\omega t)\,dt$$

$$= \int_{-\infty}^{\infty} \exp\{-\tfrac{1}{2}(t^2 + 2j\omega t)\}\,dt$$

$$= \int_{-\infty}^{\infty} \exp(-\tfrac{1}{2}\omega^2)\exp\{-\tfrac{1}{2}(t^2 + 2j\omega t - \omega^2)\}\,dt$$

$$= \exp(-\tfrac{1}{2}\omega^2)\int_{-\infty}^{\infty} \exp\{-\tfrac{1}{2}(t + j\omega)^2\}\,dt$$

$$= \sqrt{2}\exp(-\tfrac{1}{2}\omega^2)\int_{-\infty}^{\infty} \exp\{-\tfrac{1}{2}(t + j\omega)^2\}\,d\{(t + j\omega)/\sqrt{2}\}$$

But

$$\int_{-\infty}^{\infty} \exp(-x^2)\,dx = \sqrt{\pi}$$

$$\therefore F(\omega) = \sqrt{(2\pi)}\exp(-\tfrac{1}{2}\omega^2)$$

Thus the Gaussian function has the same form, viz. $\exp(-\frac{1}{2}x^2)$ in both the time and frequency representations (cf. the infinite train of narrow pulses considered in section 2.6) and is said to be self-reciprocal.

It is interesting to note that the Gaussian function is one of a set known as the Hermite functions. These are given by the relation

$$H_n(t) = \frac{d^n}{dt^n}\exp(-\tfrac{1}{2}t^2)$$

Thus the Gaussian function is $H_0(t)$. All the Hermite functions have the property of self-reciprocity.

4.8 The Unit Step Function

The unit step function is defined by the relations

$$f(t) = 1 \quad t > 0$$
$$= 0 \quad t < 0$$

and is shown in Fig. 4.4. This is another important function which finds many applications in engineering, e.g. in circuit theory. We note

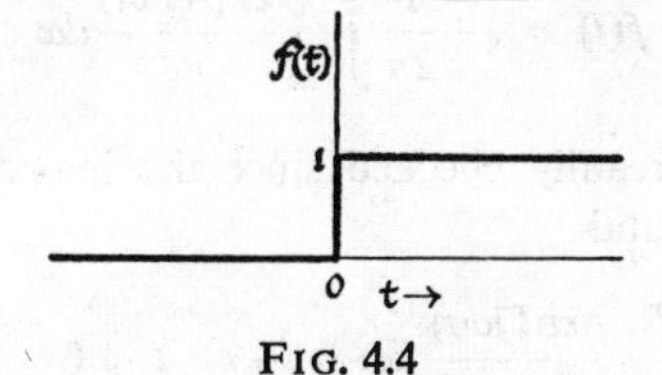

FIG. 4.4
The unit step function

that as defined the unit step function does not satisfy Dirichlet's conditions since $\int_{-\infty}^{\infty} |f(t)|\,dt$ is not finite. To overcome this difficulty we define a new function

$$f(t) = 1 \quad 0 < t < T$$
$$= 0 \quad \text{elsewhere}$$

Dirichlet's conditions are now satisfied and we can proceed with this function. The spectrum is given by

$$F(\omega) = \int_{-\infty}^{\infty} f(t) \exp(-j\omega t)\,dt = \int_0^T \exp(-j\omega t)\,dt$$

$$= \frac{1}{j\omega}\{1-\exp(-j\omega T)\}$$

Hence we have

$$f(t) = \frac{1}{2\pi}\int_{-\infty}^{\infty} \frac{1}{j\omega}\{1-\exp(-j\omega T)\}\exp(j\omega t)\,d\omega$$

$$= \frac{1}{2\pi}\int_{-\infty}^{\infty} \frac{1}{j\omega}\exp(j\omega t)\,d\omega - \frac{1}{2\pi}\int_{-\infty}^{\infty} \frac{1}{j\omega}\exp\{j\omega(t-T)\}\,d\omega$$

We can make this function approximate the unit step function by making T large and under these conditions

$$\int_{-\infty}^{\infty} \frac{1}{j\omega}\exp\{j\omega(t-T)\}\,d\omega = \int_{-\infty}^{\infty} \frac{\exp(-j\omega T)}{j\omega}\,d\omega$$

which is a standard integral and has value $-\pi$ for $T > 0$.

$$f(t) = \tfrac{1}{2} + \frac{1}{2\pi}\int_{-\infty}^{\infty} \frac{\exp(j\omega t)}{j\omega}\,d\omega$$

This result is readily checked since the integral is similarly a standard integral and

$$\int_{-\infty}^{\infty} \frac{\exp(j\omega t)}{j\omega}\,d\omega = -\pi \quad t < 0$$

$$= +\pi \quad t > 0$$

Hence

$$f(t) = 1 \quad t > 0$$

$$= 0 \quad t < 0$$

as required.

From the formula

$$f(t) = \tfrac{1}{2} + \frac{1}{2\pi}\int_{-\infty}^{\infty}\left(\frac{1}{j\omega}\right)\exp(j\omega t)\,d\omega$$

we see that there is a d.c. term of magnitude $\frac{1}{2}$ which has a spectrum consisting of an impulse of magnitude π situated at $\omega=0$ (see section 4.5). The spectrum of the second term in the expression is clearly $(1/j\omega)$ and so from the superposition theorem the complete spectrum of the unit step function is an impulse of magnitude π at $\omega=0$ plus the function $F(\omega)=(1/j\omega)$. Hence the magnitude of the component frequencies in the unit step function is inversely proportional to the frequency.

4.9 The Exponential Decay

This is another function which frequently arises in engineering practice. It is defined by

$$\begin{aligned} f(t) &= \exp(-\alpha t) \quad t > 0 \\ &= 0 \quad\quad\quad\quad\; t < 0 \end{aligned}$$

A sketch of the function is shown in Fig. 4.5. The determination of the spectrum is quite straightforward.

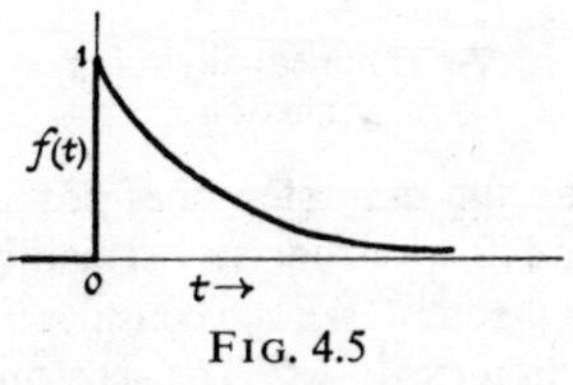

FIG. 4.5
An exponential decay

$$\begin{aligned} F(\omega) &= \int_{-\infty}^{\infty} f(t)\exp(-j\omega t)\,dt \\ &= \int_{0}^{\infty}\exp(-\alpha t)\exp(-j\omega t)\,dt = \int_{0}^{\infty}\exp\{-(\alpha+j\omega)\,t\}\,dt \\ &= \frac{-1}{\alpha+j\omega}[\exp\{-(\alpha+j\omega)\,t\}]_{0}^{\infty} = \frac{1}{\alpha+j\omega} \end{aligned}$$

Thus

$$F(\omega) = \frac{1}{\alpha + j\omega}$$

$$|F(\omega)| = \sqrt{[F(\omega)\,F^*(\omega)]} = \sqrt{\left[\frac{1}{(\alpha+j\omega)(\alpha-j\omega)}\right]}$$

$$\therefore |F(\omega)| = \frac{1}{\sqrt{(\alpha^2+\omega^2)}}$$

4.10 The Exponential Rise

This function is related to the previous one and they often occur together in different parts of the same system (e.g. the transient response of an RC circuit). It is sketched in Fig. 4.6 and defined by

$$\begin{aligned} f(t) &= 1 - \exp(-\alpha t) \quad & t > 0 \\ &= 0 & t < 0 \end{aligned}$$

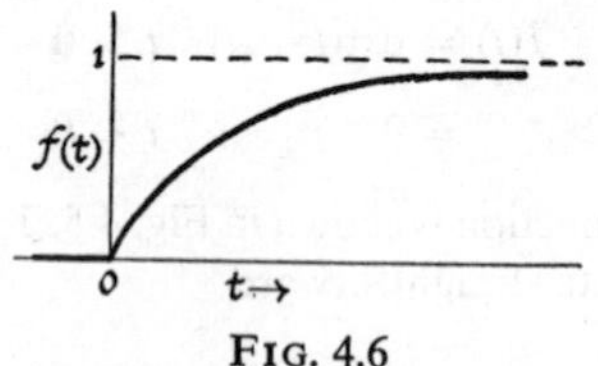

FIG. 4.6
An exponentially rising function

This function, like the unit, step does not satisfy the condition $\int_{-\infty}^{\infty} |f(t)|\, dt$ finite and its spectrum is most readily obtained by means of the superposition theorem. We can consider this function as a unit step minus the function $\exp(-\alpha t)$. The spectrum of the unit step is given by $F(\omega) = 1/j\omega$ and the formula for $f(t)$ is

$$f(t) = \tfrac{1}{2} + \frac{1}{2\pi}\int_{-\infty}^{\infty} \frac{1}{j\omega}\exp(j\omega t)\, d\omega$$

Hence for the exponential rise the spectrum is given by

$$F(\omega) = \frac{1}{j\omega} - \frac{1}{\alpha + j\omega} = \frac{\alpha}{j\omega(\alpha + j\omega)}$$

and

$$|F(\omega)| = \sqrt{\left[\frac{-j\alpha}{\omega(\alpha+j\omega)}\cdot\frac{j\alpha}{\omega(\alpha-j\omega)}\right]}$$

$$\therefore\ |F(\omega)| = \frac{\alpha}{\omega\sqrt{(\alpha^2+\omega^2)}}$$

and

$$f(t) = \tfrac{1}{2}+\frac{1}{2\pi}\int_{-\infty}^{\infty}\frac{\alpha}{j\omega(\alpha+j\omega)}\exp(j\omega t)\,d\omega$$

4.11 The Cosine Wave of Finite Duration

This is obviously another case which arises in practice since it represents a co-sinusoidal oscillation which is switched on at some point in time and then off again at a later time. The r.f. pulse commonly used in radar engineering is just such a waveform. A sketch of the function is shown in Fig. 4.7. This function is defined by

$$\begin{aligned} f(t) &= \cos pt \quad -\tfrac{1}{2}T < t < \tfrac{1}{2}T \\ &= 0 \qquad \text{elsewhere} \end{aligned}$$

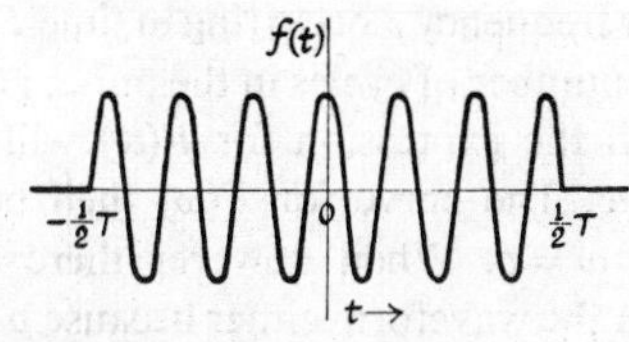

FIG. 4.7
A cosine wave of finite duration

We obtain the spectrum as

$$F(\omega) = \int_{-\infty}^{\infty} f(t)\exp(-j\omega t)\,dt = \int_{-T/2}^{T/2}\cos pt\exp(-j\omega t)\,dt$$

Whilst this is a standard integral and the integration can quite readily be carried through it is a little lengthy to do. A simpler integral results if we make use of the properties of $F(\omega)$ which we have deduced. The function $f(t)$ in this case is even and therefore its Fourier transform is a pure real function, i.e. it is equal to $a(\omega)$ the cosine transform (section 3.4). Hence

$$F(\omega) = a(\omega) = \int_{-\infty}^{\infty} f(t) \cos \omega t \, dt$$

$$\therefore F(\omega) = \int_{-T/2}^{T/2} \cos \omega t \cos pt \, dt$$

$$= \tfrac{1}{2} \int_{-T/2}^{T/2} \{\cos (\omega - p) t + \cos (\omega + p) t\} \, dt$$

$$= \tfrac{1}{2} \left[\frac{1}{\omega - p} \sin (\omega - p) t + \frac{1}{\omega + p} \sin (\omega + p) t \right]_{-T/2}^{T/2}$$

$$= \frac{1}{\omega - p} \sin (\omega - p) \frac{T}{2} + \frac{1}{\omega + p} \sin (\omega + p) \frac{T}{2}$$

which may be written

$$F(\omega) = \frac{T}{2} \left[\frac{\sin \frac{1}{2}(\omega - p) T}{\frac{1}{2}(\omega - p) T} + \frac{\sin \frac{1}{2}(\omega + p) T}{\frac{1}{2}(\omega + p) T} \right]$$

The number of cycles of the waveform in the pulse, i.e. the number of cycles of angular frequency p occurring in time T is given by $pT/2\pi$. If there are a large number of cycles in the pulse, i.e. pT is large, then the second term in the expression for $F(\omega)$ will be small for the positive range of ω. The curve for $F(\omega)$ then becomes the curve $\sin x/x$ centred at $\omega = p$. When however there are only a small number of cycles in the waveform either because p is small, i.e. a low frequency, or because T is small, i.e. a short pulse, then the second term makes an important contribution to $F(\omega)$ and the curve changes its shape.

Consider for example the case shown in Fig. 4.8(*a*) where there is only one half cycle in the waveform. In this case $pT = \pi$. Hence we have

$$F(\omega) = \frac{1}{\omega - p} \sin \left(\frac{\pi \omega}{2p} - \frac{\pi}{2} \right) + \frac{1}{\omega + p} \sin \left(\frac{\pi \omega}{2p} + \frac{\pi}{2} \right)$$

$$\therefore F(\omega) = \frac{-2p}{\omega^2 - p^2} \cos \frac{\pi \omega}{2p}$$

and on writing $\omega/p = x$ this becomes

$$F(\omega) = \frac{2}{p(1-x^2)} \cos \tfrac{1}{2}\pi x$$

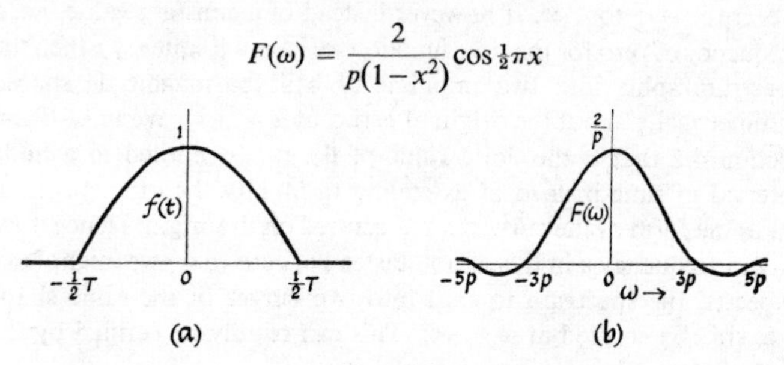

FIG. 4.8
(*a*) A cosine wave of half period duration and (*b*) its spectrum

We can sketch this curve as follows. For $\omega = 0$, i.e. $x = 0$, we have $F(0) = 2/p$. At the point $\omega = p$, or $x = 1$, both numerator and denominator become zero and we evaluate the expression by differentiating numerator and denominator and then putting $x = 1$. Thus

$$F(1) = \left[\frac{2}{p}\frac{\pi}{2}\frac{\sin \frac{1}{2}\pi x}{2x}\right]_{x=1} = \frac{\pi}{2p}$$

The expression for $F(\omega)$ has zeros where $\cos(\frac{1}{2}\pi x) = 0$, i.e. where $\frac{1}{2}\pi x = \frac{1}{2}(2n+1)\pi$, or $\omega = (2n+1)p$ (except the point $\omega = p$ which we have already dealt with. When x becomes large $\cos(\frac{1}{2}\pi x)$ oscillates between $+1$ and -1, and $(1-x^2)$ increases in magnitude continuously. Hence the magnitude of the oscillations of the expression for $F(\omega)$ decreases continuously as ω increases. The curve is sketched in Fig. 4.8(*b*).

It is instructive to examine this change of shape a little more closely We derived the expression for $F(\omega)$

$$F(\omega) = \frac{T}{2}\left[\frac{\sin \frac{1}{2}(\omega-p)T}{\frac{1}{2}(\omega-p)T} + \frac{\sin \frac{1}{2}(\omega+p)T}{\frac{1}{2}(\omega+p)T}\right]$$

Thus the expression consists of the sum of two curves of the form $\sin x/x$, one centred at $\omega = +p$ and one at $\omega = -p$. We might have expected this from our investigation of the impulse in section 4.5.

We saw there that an impulse at the origin represents a constant d.c. from $-\infty$ to $+\infty$. If however instead of a constant value, i.e. a frequency of zero for the time function we have a frequency p then the spectrum splits into two impulses of half the magnitude spaced symmetrically about the origin of ω (i.e. at $\omega = \pm p$). We know from section 4.2 that if the finite value of the d.c. is limited to a finite interval of time instead of extending to infinity then the spectrum takes the form of the curve $\sin x/x$ centred on the origin. Hence if we have a frequency p in this region instead of zero (d.c.) we might have expected the spectrum to split into two curves of the same shape (i.e. $\sin x/x$) centred at $\omega = \pm p$. This can readily be verified by the shift theorem. Let $f(t)$ be defined by

$$\begin{aligned} f(t) &= 1 \quad -\tfrac{1}{2}T < t < \tfrac{1}{2}T \\ &= 0 \quad \text{elsewhere} \end{aligned}$$

Then from section 4.2

$$F(\omega) = T\frac{\sin \frac{1}{2}\omega T}{\frac{1}{2}\omega T}$$

If now we shift this curve to $\omega = p$ and take half the magnitude then we have

$$F(\omega - p) = \frac{T}{2}\frac{\sin \frac{1}{2}(\omega-p)T}{\frac{1}{2}(\omega-p)T}$$

By the second shift theorem the signal of this spectrum is given by $\frac{1}{2}\exp(jpt).f(t)$. Shifting the curve to $\omega = -p$ gives

$$F(\omega+p) = \frac{T}{2}\left[\frac{\sin \frac{1}{2}(\omega+p)T}{\frac{1}{2}(\omega+p)T}\right]$$

which by the shift theorem has signal $\frac{1}{2}\exp(-jpt).f(t)$. Hence by the superposition theorem the signal due to a spectrum

$$F(\omega) = \frac{T}{2}\left[\frac{\sin \frac{1}{2}(\omega-p)T}{\frac{1}{2}(\omega-p)T} + \frac{\sin \frac{1}{2}(\omega+p)T}{\frac{1}{2}(\omega+p)T}\right]$$

is given by

$$\begin{aligned} &\tfrac{1}{2}\{\exp(jpt)+\exp(-jpt)\}f(t) \\ &= \cos pt.f(t) = \cos pt \quad -\tfrac{1}{2}T < t < \tfrac{1}{2}T \\ &\qquad\qquad\qquad = 0 \qquad \text{elsewhere} \end{aligned}$$

We can use these ideas to explain the change of shape of the curve noted above. When there are a large number of cycles in the waveform pT is large and the two branches of the curve of $F(\omega)$ are independent. The one centred at $\omega = -p$ becomes negligibly small over the region where the one centred at $\omega = +p$ is important as shown in Fig. 4.9(*a*).

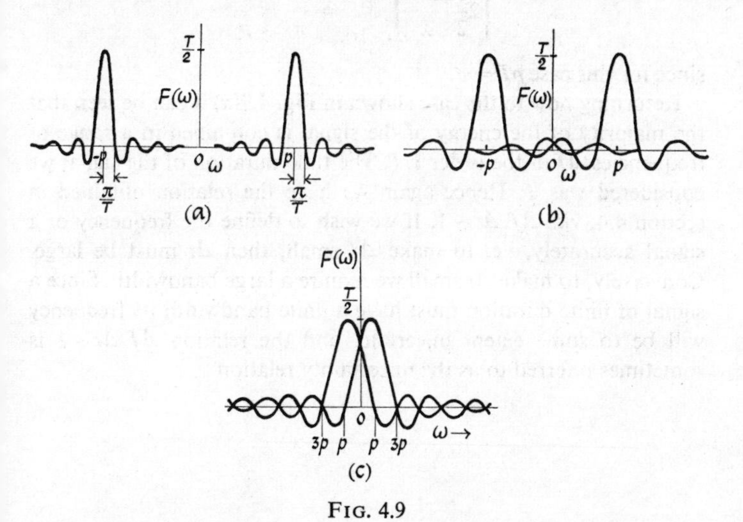

FIG. 4.9

Change of spectrum as the duration of the cosine wave varies: (*a*) the positive and negative parts are isolated from each other; (*b*) some overlap occurs; and (*c*) the case for half period duration

If the time T is reduced keeping p constant then the positions of the first minima move away from the point at which the curve is centred and the curves are stretched out in the t direction. If p is reduced keeping T constant then the two branches of the curve move towards each other keeping the same shape. In both cases a point is eventually reached when the two curves begin to overlap and the curve centred at $\omega = -p$ still has an appreciable value over part of the region where the curve centred at $\omega = +p$ is important. This condition is shown in Fig. 4.9(*b*). Since the curve for $F(\omega)$ is the sum of the two branches then at the point where the overlap becomes important the curve begins to change shape. The condition $pT = \pi$, the case considered above, is shown in Fig. 4.9(*c*). In this case the two curves combine in such a way as to give a resultant curve with one main peak at the origin. From the figure we see that the value of this peak is given by

$$\left[2\frac{T}{2}\frac{\sin x}{x}\right]_{x=\frac{1}{2}\pi} = \frac{2T}{\pi} = \frac{2}{p}$$

since for this case $pT = \pi$.

Returning now to the case shown in Fig. 4.9(*a*) it can be seen that the majority of the energy of the signal is contained in a range of frequencies Δf of the order $1/T$. The time duration of the signal we considered was T. Hence again we have the relation obtained in section 4.4, viz. $\Delta f . \Delta t \sim 1$. If we wish to define the frequency of a signal accurately, i.e. to make Δf small, then Δt must be large. Conversely, to make Δt small we require a large bandwidth. Since a signal of finite duration must have a finite bandwidth its frequency will be to some extent uncertain, and the relation $\Delta f . \Delta t \sim 1$ is sometimes referred to as the uncertainty relation.

CHAPTER V

Application to Circuit Analysis

5.1 Introduction

In this chapter we shall apply the ideas developed earlier in the book to some of the problems which arise in dealing with electrical circuits.

5.2 Filters

We shall consider in this section only linear filters, i.e. those filters whose output is proportional to the input. The factor of proportionality however is not the same at all frequencies, and this gives rise to the filtering action. The filter may be specified in terms of its frequency characteristic or transfer function as it is sometimes termed. Thus if we represent the transfer function by $Z_T(\omega)$ then the output at frequency ω_0 is $Z_T(\omega_0)$ times the input. Hence if the input is represented as a function of frequency, i.e. as a spectrum $F_i(\omega)$, then the output represented as a function of frequency is

$$F_o(\omega) = Z_T(\omega) F_i(\omega)$$

If, then, we have a waveform $f_i(t)$ as the input to a filter we can obtain the output by the use of Fourier transforms. We can write each of the complex functions in the above equation in terms of amplitude and phase. Hence we have

$$R_o(\omega) \exp\{j\theta_o(\omega)\} = A(\omega) \exp\{j\phi(\omega)\} R_i(\omega) \exp\{j\theta_i(\omega)\}$$

where $A(\omega)$ is the amplitude characteristic of the filter and $\phi(\omega)$ its phase characteristic.

It is convenient to consider the general case where ω is allowed to take negative as well as positive values and we shall now deduce some of the properties of $Z_T(\omega)$. We know that

$$F(-\omega) = F^*(\omega)$$

and so it follows immediately that

$$Z_T(-\omega) = Z_T^*(\omega)$$

and hence $$A(-\omega) = A(\omega)$$

and $$\phi(-\omega) = -\phi(\omega)$$

The amplitude characteristic is thus an even or symmetric function of ω and the phase characteristic is odd or anti-symmetric.

5.3 Distortionless Transmission

If a waveform is to be transmitted without distortion then the only operations which can be carried out upon it are to multiply it by a constant and to shift it in time. These operations correspond to amplification or attenuation, and delay. If we multiply a time function $f(t)$ by a constant A then all the frequency components are also multiplied by A. If the function is shifted in time, i.e. it becomes $f(t-T)$, then the frequency representation is multiplied by $\exp(-j\omega T)$, by the shift theorem. Thus the general characteristic of a filter which will transmit a waveform without distortion is

$$Z_T(\omega) = A \exp(-j\omega T)$$

This is readily verified, for if the input wave is given by $f_i(t)$ then

$$f_i(t) = \frac{1}{2\pi}\int_{-\infty}^{\infty} F_i(\omega)\exp(j\omega t)\,d\omega$$

and

$$F_o(\omega) = Z_T(\omega)F_i(\omega) = A\exp(-j\omega T)F_i(\omega)$$

so that

$$f_o(t) = \frac{1}{2\pi}\int_{-\infty}^{\infty} F_o(\omega)\exp(j\omega t)\,d\omega$$

$$= \frac{1}{2\pi}\int_{-\infty}^{\infty} \{A\exp(-j\omega T)F_i(\omega)\}\exp(j\omega t)\,d\omega$$

or $$f_o(t) = Af_i(t-T)$$

We can illustrate this by considering a transmission line. If $V(x,t)$ is the voltage at a point distant x from the input at time t then it may be shown that for a sinusoidal input of frequency ω_0

$$V(x,t) = \mathscr{V}\exp j(\omega_0 t - Px)$$

where $\mathscr{V}$ is the amplitude of the voltage at the input and P is the propagation constant. The input voltage is given by $V_i(t) = V(0,t)$ and the output voltage $V_o(t) = V(X,t)$, where X is the length of the transmission line. Thus we have

$$V_i(t) = \mathscr{V} \exp(j\omega_0 t)$$

and

$$V_o(t) = \mathscr{V} \exp j(\omega_0 t - PX) = V_i(t) \exp(-jPX)$$

Thus at the frequency ω_0 the output voltage is $\exp(-jPX)$ times the input where P the propagation constant will be a function of frequency. The magnitude of the frequency components at the output is obtained by multiplying those at the input by $\exp\{-jP(\omega)X\}$. This function then is the transfer impedance of the transmission line. It may be shown that the function $P(\omega)$ is given by

$$P(\omega) = \pm\sqrt{\{-(G+j\omega C)(R+j\omega L)\}}$$

where G is the shunt conductance, C is the capacitance, R is the series resistance and L is the inductance per unit length of the transmission line. Hence we have

$$Z_T(\omega) = \exp[\pm jX\sqrt{\{-(G+j\omega C)(R+j\omega L)\}}]$$

For distortionless transmission the transfer function must be of the form $A\exp(-j\omega T)$ and we can show that this form is obtained by making

$$(G+j\omega C) = \mu^2(R+j\omega L)$$

where μ is a constant.

If this relation is satisfied then

$$Z_T(\omega) = \exp[\pm jX\sqrt{\{-\mu^2(R+j\omega L)^2\}}]$$

$$= \exp[\pm \mu X(R+j\omega L)]$$

$$\therefore Z_T(\omega) = \exp(\pm \mu XR) \exp(\pm j\mu X\omega L)$$

We note from above that μ has dimensions $[R]^{-1}$ and that R and L are measured per unit length. Thus μXR is dimensionless and constant. Considering for example the negative signs we can write $\exp(-\mu XR) = A$. Similarly it can be seen that $\mu X\omega L$ is dimensionless and can therefore be written ωT where T is the time given by μXL and is constant. Hence

$$Z_T(\omega) = A \exp(-j\omega T)$$

which is of the form required. Thus a transmission line will transmit signals without distortion if its parameters are related by the formula

$$(G+j\omega C) = \mu^2(R+j\omega L)$$

a well-known result in transmission line theory.

5.4 Passage of a Step Function through an Ideal Filter

If a filter does not have a transfer characteristic of the form developed in the last section then, in general, distortion will result. For example, if A is a function of frequency so that some frequencies are attenuated more than others then the function of time appearing at the output will be different from that at the input. We shall investigate this

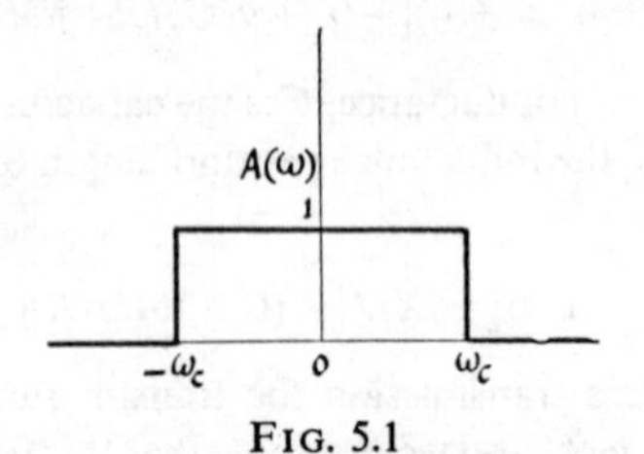

FIG. 5.1
A perfect filter characteristic

phenomenon by considering the passage of a step function through an ideal filter. We suppose that the filter passes all frequencies up to frequency ω_c without modification and completely stops all those above ω_c. The transfer characteristic is therefore as shown in Fig. 5.1. This may be written

$$\begin{aligned} A(\omega) &= 1 \quad -\omega_c < \omega < \omega_c \\ &= 0 \quad \text{elsewhere} \end{aligned}$$

Consider the symmetrical step function given by

$$\begin{aligned} f_i(t) &= -\tfrac{1}{2} \quad t < 0 \\ &= +\tfrac{1}{2} \quad t > 0 \end{aligned}$$

Then from section 4.8

$$F_i(\omega) = \frac{1}{j\omega}$$

Hence

$$F_o(\omega) = Z_T(\omega)\,F_i(\omega) = \frac{1}{j\omega} \qquad -\omega_c < \omega < \omega_c$$

$$= 0 \qquad \text{elsewhere}$$

so that

$$f_o(t) = \frac{1}{2\pi}\int_{-\omega_c}^{\omega_c} \frac{1}{j\omega}\exp(j\omega t)\,d\omega$$

$$= \frac{1}{2\pi}\int_{-\omega_c}^{\omega_c} \frac{\cos\omega t}{j\omega}\,d\omega + \frac{1}{2\pi}\int_{-\omega_c}^{\omega_c} \frac{j\sin\omega t}{j\omega}\,d\omega$$

$$= \frac{1}{\pi}\int_0^{\omega_c} \frac{\sin\omega t}{\omega}\,d\omega$$

since the first integral has an odd integrand in ω and the second has an even one. Putting $k = \omega t$

$$f_o(t) = \frac{1}{\pi}\int_0^{\omega_c t} \frac{\sin k}{k}\,dk$$

$$\therefore f_o(t) = \frac{1}{\pi}\,Si(\omega_c t)$$

where $Si(\omega_c t)$ is the sine integral which we discussed in section 2.8. If we had considered the unit step function then we should have had the relation (see section 4.8)

$$f_i(t) = \tfrac{1}{2} + \frac{1}{2\pi}\int_{-\infty}^{\infty} \frac{\exp(j\omega t)}{j\omega}\,d\omega$$

Hence the output function $f_o(t)$ would be as shown in Fig. 5.2.

It is seen that the output voltage overshoots the final steady value and oscillates about it, the amplitude of the oscillations decreasing with time. As ω_c increases the value of the overshoot is not altered but the time at which the maximum occurs is smaller, the whole curve being compressed towards the origin but remaining unaltered in the y direction (cf. Gibbs' phenomenon, section 2.8).

It is also seen that there is in addition to the overshoot what has been termed an anticipatory transient. Although the step function does not occur until the time $t = 0$ there are oscillations in the output waveform before this instant. We therefore have an output with no input in the region $t < 0$—the output anticipates the input. In the above analysis we neglected the phase characteristic of the filter. This as we

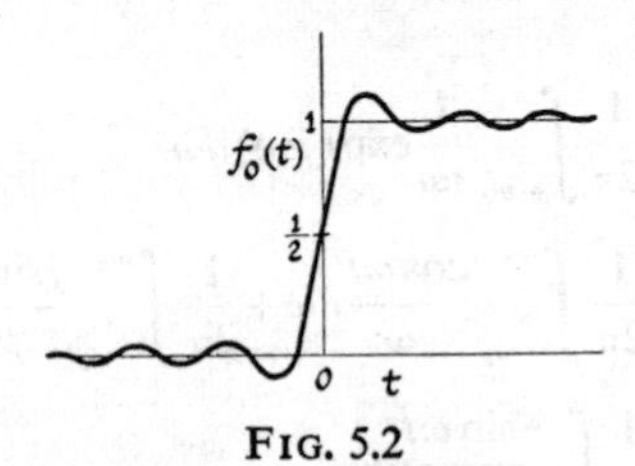

FIG. 5.2
The output of a perfect filter for a step function input at $t=0$

have seen introduces delay. Hence taking this into account the whole curve of Fig. 5.2 should be shifted to the right. However the curve of Fig. 5.2 will not become zero (except at discrete points) until t becomes infinite and hence there is always a small anticipatory transient for a finite delay. A filter with the characteristics assumed is not physically realizable.

5.5 Passage of a Step Function through a CR Circuit

We have developed a method of calculating the output of a filter by first obtaining the spectrum of the output and then taking the inverse Fourier transform. In the last section we obtained an output which extended to infinity in both positive and negative directions. Its spectrum however was confined to a finite region of the frequency spectrum by virtue of the action of the filter assumed. This is an example of the phenomenon of reciprocal spreading which we discussed in section 4.4. If the function is limited in one representation it spreads in the other. Thus to obtain a function of time which is confined to a limited interval of time, say $t > 0$, corresponding to a physical case, then the frequency representation must extend to infinity and can have no sharp cut-off of the type assumed in the last section. We have already calculated the Fourier transforms of several functions which have a discontinuity at the origin, i.e. functions which start at some definite point in time, a typical one being the exponential decay of section 4.9 which has spectrum $1/(\alpha + j\omega)$. It is interesting to work through the inverse process to illustrate how spectra can be formed by filters which lead to time functions with a discontinuity at the origin.

Consider for example the *CR* circuit shown in Fig. 5.3. The transfer function of this circuit is given by

$$Z_T(\omega) = \frac{R}{R + (1/j\omega C)} = \frac{j\omega}{\alpha + j\omega}$$

where $\alpha = 1/CR$.

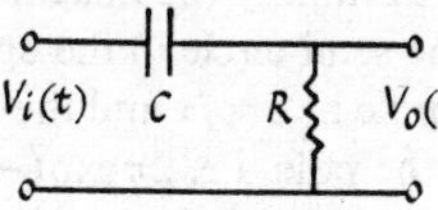

$V_i(t)$ C R $V_o(t)$

Fig. 5.3
The *CR* circuit

If the input $V_i(t)$ is a step function then the output spectrum is given by

$$F_o(\omega) = \frac{1}{j\omega}\frac{j\omega}{\alpha + j\omega} = \frac{1}{\alpha + j\omega}$$

We already know from the results of section 4.9 that the signal of this spectrum is given by

$$\begin{aligned} f(t) &= \exp(-\alpha t) \quad t > 0 \\ &= 0 \quad t < 0 \end{aligned}$$

and the output is therefore an exponential decay, a well-known result in circuit theory. It is interesting however to see how this discontinuity arises from the inverse transform. The output waveform is given by

$$f_o(t) = \frac{1}{2\pi}\int_{-\infty}^{\infty} \frac{\exp(j\omega t)}{\alpha + j\omega}\, d\omega$$

This integral can be evaluated by the contour integration method. Consider the integral

$$f(z) = \int_C \frac{\exp(jtz)}{\alpha + jz} dz$$

where C is a contour in the complex z plane. This type of integral is evaluated in terms of the residues at the poles, and in this case there is one pole at $z = j\alpha$, which has residue $\exp(-\alpha t)$. The contour C must be chosen so that the integral converges. In this case the infinite semi-circles on the real axis are appropriate, and which of these two is used depends upon the parameter t. If t is negative then for the integral to converge on the semi-circle at infinity the imaginary part of z must be negative and the semi-circle in the lower half plane must be used. This does not enclose any singularities of the function and hence the integral is zero. For t positive, then for the integral to converge on the semi-circle at infinity the imaginary part of z must also be positive and hence the semi-circle in the upper half plane is to be used. This encloses the pole at $z = j\alpha$ and the value of the integral is 2π times the residue at the pole, i.e. $2\pi \exp(-\alpha t)$. Thus the integral gives rise to a function of t which is zero for negative t and has a discontinuity at the origin. On completing the calculation the result is obtained

$$\begin{aligned} f_o(t) &= \exp(-\alpha t) \quad & t > 0 \\ &= 0 & t < 0 \end{aligned}$$

in agreement with that obtained earlier. (For the details of carrying out this type of integration see, for example, Phillips (1943).)

The transient response of circuits (the output when the input is a step function or an impulse function) is not usually calculated by means of Fourier transforms but by the use of another integral transform, the Laplace transform. In this case the kernel

$$K(p, t) = \exp(-pt)$$

and the limits are from 0 to ∞. Thus the Laplace transform is defined by

$$\mathscr{L}(p) = \int_0^\infty f(t) \exp(-pt)\, dt$$

where p is in general complex. The exponential function now has a real part as well as an imaginary part and problems of convergence are less likely to be met with. This has proved a powerful tool in many applications and there are many books on the subject (see, for example, Jaeger (1961)).

5.6 Initial Rate of Rise and Bandwidth

We noted in section 5.4 that as the value of the cut-off frequency ω_c of the filter increases the curve for the time function is compressed towards the origin. Thus the rate of rise at the change-over increases. We can relate the initial rate of rise to the bandwidth ω_c as follows. We have for the output function

$$f_o(t) = \frac{1}{\pi}\int_0^{\omega_c t} \frac{\sin k}{k}\,dk$$

Now it is shown in the appendix that

$$\frac{d}{dt}\int_0^{at} g(x)\,dx = ag(at)$$

Hence using this formula

$$\frac{d}{dt}f_o(t) = \frac{\omega_c}{\pi}\frac{\sin \omega_c t}{\omega_c t}$$

This function is a maximum at $t = 0$ and hence the initial rate of rise is the maximum rate of rise also, and is given by

$$\left[\frac{d}{dt}f_o(t)\right]_{t=0} = \frac{\omega_c}{\pi}$$

We see that the rate of rise is proportional to the bandwidth and so the more rapid the rate of change in the signals which are to be transmitted, the wider the bandwidth of the filters must be.

We can also calculate quite simply the value of the initial rate of rise for the CR circuit. Here we have

$$f_o(t) = 1-\exp(-\alpha t)$$

$$\therefore \left[\frac{d}{dt}f_o(t)\right]_{t=0} = \alpha = \frac{1}{CR}$$

The filter characteristic is given by

$$Z_T(\omega) = \frac{1}{1+j\omega CR}$$

In this case the definition of bandwidth is somewhat arbitrary. If we take the '6db point', i.e. where $|Z_T(\omega)| = \frac{1}{2}$ then $\omega_c CR = \sqrt{3}$ or $\omega_c = \sqrt{3}/(CR)$

$$\therefore \left[\frac{d}{dt} f_o(t)\right]_{t=0} = \frac{1}{CR} = \frac{\omega_c}{\sqrt{3}}$$

This is a similar result to that obtained above for the ideal filter. Numerical agreement could not have been expected owing to the arbitrary definition of bandwidth ω_c in the second case.

The foregoing discussion shows that the bandwidth required in a transmission system is closely related to the rise times it must transmit. The above analysis is readily extended to the case of pulses and here the rise times will determine how accurately the pulse shapes are preserved and also set the limit on the ability of the system to resolve consecutive pulses. It may be shown that two pulses separated by a time Δt will just be resolved by a filter with a bandwidth Δf if Δf and Δt are related by the uncertainty relation $\Delta f . \Delta t \sim 1$ (see also sections 4.4 and 4.11). Alternatively, the bandwidth required in the system can be obtained from a consideration of the spectrum of the pulses. A detailed discussion of these considerations will be found in Goldman (1948).

5.7 Response of a Circuit to an Input Function

Consider first the case where the input function is a unit impulse function $f_i(t)$. From the results of section 4.5 we have that the input spectrum $F_i(\omega) = 1$. Hence for a network with a transfer function $Z_T(\omega)$ we have for the output spectrum $F_o(\omega)$

$$F_o(\omega) = Z_T(\omega) F_i(\omega) = Z_T(\omega)$$

and the output signal $f_o(t)$

$$f_o(t) = \frac{1}{2\pi} \int_{-\infty}^{\infty} Z_T(\omega) \exp(j\omega t)\, d\omega$$

and so

$$Z_T(\omega) = \int_{-\infty}^{\infty} f_o(t) \exp(-j\omega t)\, dt$$

Thus we have the important result that the transfer function of a network and its impulse response form a Fourier pair.

The impulse response $f_0(t)$ of the network is the output when a unit impulse is applied at time $t=0$. If the impulse is applied at time $t=\tau$ the output is $f_0(t-\tau)$. Further if the impulse is not of unit magnitude but of magnitude $f_1(\tau)$ then the output is given by $f_1(\tau)\, f_0(t-\tau)$. Any arbitrary input function can be made up of a series of impulses of the appropriate magnitude and the output due to an input function $f_1(\tau)$ is the integral of the outputs due to inputs consisting of impulses of magnitude $f_1(\tau)$ occurring at time τ. Hence the output function $f_2(t)$ is given by

$$f_2(t) = \int_{-\infty}^{\infty} f_1(\tau) f_0(t-\tau)\, d\tau = f_1(t) * f_0(t).$$

This is the convolution integral. The output signal is therefore the convolution of the input signal with the impulse response of the network. The output spectrum is the product of the input spectrum and the transfer function of the network. We may compare these results with those obtained in section 3.13.

5.8 The Method of Paired Echoes

We shall now consider the effect on the output waveform of a filter which is not flat in the pass band. We have already seen that this will introduce distortion into the output and we now develop a method of calculating what this will be. In Fig. 5.4 we show the characteristic of an ideal filter by the dotted line. This has a value of one in the pass band and a sharp cut-off at $\omega_c = \pi/\tau$. We know that this characteristic must be an even function of ω so the simplest form of variation we can introduce is a cosinusoidal variation with one complete cycle in the pass band. If this has amplitude a then the filter characteristic is represented by

$$\begin{aligned} |Z_T(\omega)| &= 1 + a\cos\omega\tau \qquad -\pi/\tau < \omega < \pi/\tau \\ &= 0 \qquad \text{elsewhere} \end{aligned}$$

In order to avoid the effects of the sharp cut-off we suppose that the input signal has a spectrum whose components are negligibly small outside the pass band, i.e. for frequencies greater than π/τ. The ideal filter of Fig. 5.4 would then pass the signal without modification. Thus with this assumption any modification to the signal occurring when it is passed through the filter shown in Fig. 5.4 will be due

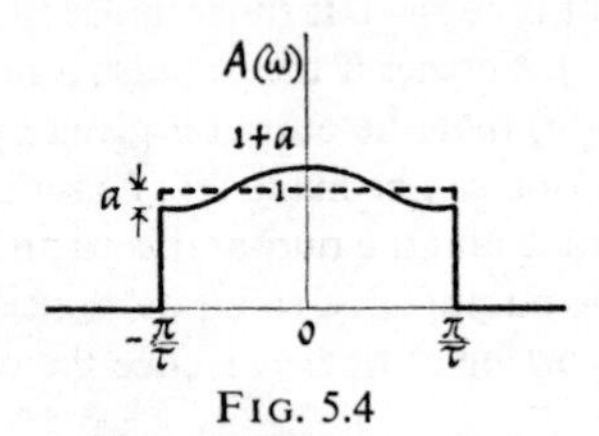

FIG. 5.4

A filter characteristic with amplitude variation in the pass band

entirely to the ripple in the pass band. We shall also assume that the phase characteristic of the filter is ideal and from section 5.3 this is given by

$$\phi(\omega) = -\omega T$$

Hence

$$Z_T(\omega) = (1+a\cos\omega\tau)\exp(-j\omega T)$$
$$= \exp(-j\omega T)+\tfrac{1}{2}a\exp\{-j\omega(T-\tau)\}+\tfrac{1}{2}a\exp\{-j\omega(T+\tau)\}$$

Therefore if $F_i(\omega)$ is the input spectrum, the output spectrum is given by

$$F_o(\omega) = F_i(\omega)[\exp(-j\omega T)+\tfrac{1}{2}a\exp\{-j\omega(T-\tau)\}$$
$$+\tfrac{1}{2}a\exp\{-j\omega(T+\tau)\}]$$

We can now apply the shift theorem to each of these three terms and if $f_i(t)$ is the input signal then we have for the output signal

$$f_o(t) = f_i(t-T)+\tfrac{1}{2}af_i(t-T+\tau)+\tfrac{1}{2}af_i(t-T-\tau)$$

This function is shown in Fig. 5.5. It is seen that in addition to the main signal which occurs after a delay T two smaller signals the same shape as $f_i(t)$ but of amplitude $\frac{1}{2}a$ occur spaced symmetrically about the main signal and at time τ from it. These are termed paired echoes. In Fig. 5.5 these are shown separated from the main signal but in some cases they would overlap. The shape of the output signal would then be different from that of the input signal but it would still be possible to resolve it into the sum of three signals, a main one and a pair of echoes. Thus the distortion introduced by a cosinusoidal variation of filter characteristic can be represented by one symmetrical set of paired echoes. It is possible to represent any variation of filter

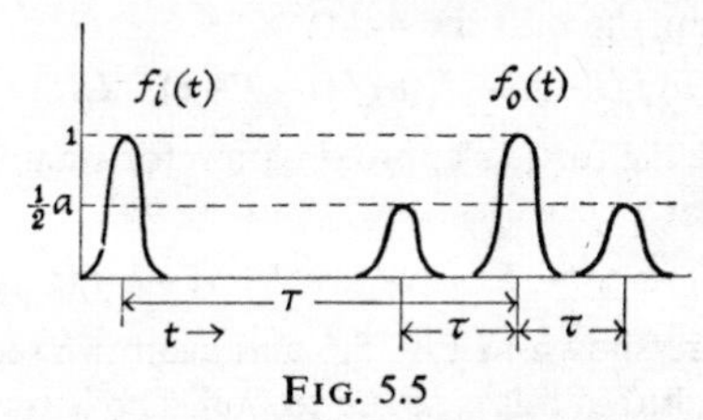

FIG. 5.5
Input and output of a filter with an imperfect amplitude characteristic

characteristic in the pass band by means of a Fourier series. Each component will give rise to a pair of echoes and the resulting distortion can therefore be represented by a set of symmetrical paired echoes.

We shall now consider the case of an imperfect phase characteristic. Since this must be an odd function of frequency it can be represented by a Fourier sine series and the simplest case is given by

$$\phi(\omega) = -\omega T + b \sin \omega\tau$$

Assuming an ideal amplitude characteristic we have

$$Z_T(\omega) = \exp\{-j(\omega T - b \sin \omega\tau)\}$$

Hence the output spectrum is given by

$$F_o(\omega) = F_i(\omega)\exp\{-j(\omega T - b \sin \omega\tau)\}$$

Now from the theory of Bessel functions we have

$$\exp\{jb \sin x\} = J_0(b) + \{\exp(jx) - \exp(-jx)\}J_1(b) + \{\exp(j2x) + \exp(-j2x)\}J_2(b) + \ldots$$

where the J's are the Bessel functions of the first kind. Suppose for simplicity that b is small so that the higher order Bessel functions can be neglected, i.e. we can write to a good approximation

$$\exp\{jb \sin \omega\tau\} = J_0(b)+\{\exp(j\omega\tau)-\exp(-j\omega\tau)\}J_1(b)$$

Then we have for the output spectrum

$$\begin{aligned} F_o(\omega) &= F_i(\omega)\exp(-j\omega T)[J_0(b)+\{\exp(j\omega\tau)-\exp(-j\omega\tau)\}J_1(b)] \\ &= J_0(b)\exp(-j\omega T)F_i(\omega)+J_1(b)\exp\{-j\omega(T-\tau)\}F_i(\omega) \\ &\quad -J_1(b)\exp\{-j\omega(T+\tau)\}F_i(\omega) \end{aligned}$$

Hence applying the shift theorem

$$f_o(t) = J_0(b)f_i(t-T)+J_1(b)f_i(t-T+\tau)-J_1(b)f_i(t-T-\tau)$$

We can make the further approximation for small $b, J_0(b) = 1$ and $J_1(b) = \frac{1}{2}b$ so that

$$f_o(t) = f_i(t-T)+\tfrac{1}{2}bf_i(t-T+\tau)-\tfrac{1}{2}bf_i(t-T-\tau)$$

The curves are shown in Fig. 5.6 and again we see that a pair of echoes appears, but in this case the second echo is reversed in amplitude. Hence a variation of the phase characteristic in the pass band

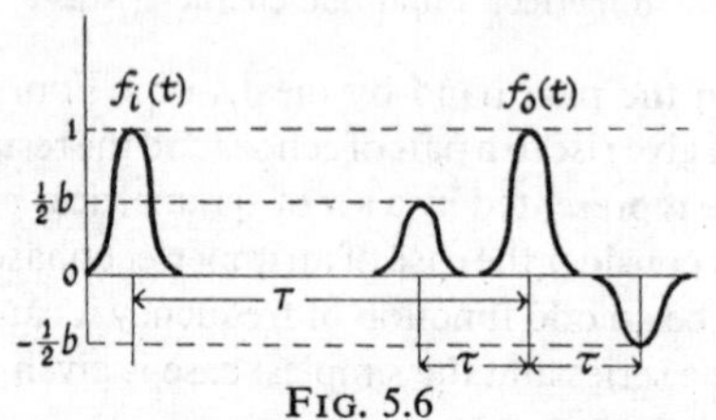

FIG. 5.6
Input and output of a filter with an imperfect phase characteristic

will give rise to distortion in the output waveform which can be represented by a set of anti-symmetrical paired echoes, the earlier one being positive and the later one negative.

REFERENCES

GOLDMAN, S. G. *Frequency Analysis, Modulation and Noise*, McGraw Hill (1948)

JAEGER, J. C. *Introduction to the Laplace Transform*, second edition, Methuen (1961)

CHAPTER VI

Application to Wave Motion Analysis

6.1 Introduction

In the previous chapters we have been dealing with the Fourier pairs $f(t)$ and $F(\omega)$ where the first is a function of time and the second a function of frequency. The use of Fourier analysis is by no means restricted to time-frequency pairs. If the functions $f(x)$ and $F(y)$ form a Fourier pair then the product (xy) must be a dimensionless quantity. We shall begin this chapter by considering the case of diffraction in which the function f will be a space variable instead of a time variable. We then pass on to consider modulation. In all cases a close parallel will be observed between the phenomena arising in diffraction theory and modulation theory simply because the underlying mathematics is the same.

6.2 Diffraction

Let us consider an absorbing screen AB as shown in Fig. 6.1 whose transmission coefficient at the point x is given by $f(x)$. Suppose the

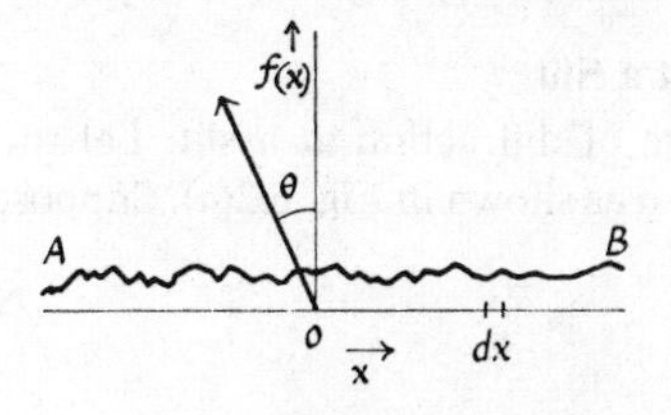

FIG. 6.1
Diffraction by a screen with variable absorption

screen is illuminated with monochromatic radiation of wavelength λ and consider the amplitude of the resulting wave in the direction θ.

The contribution of the element dx at the point x has amplitude proportional to $f(x)$ and phase $(2\pi \sin \theta/\lambda)x$, or Kx where

$$K = (2\pi \sin \theta/\lambda)$$

Therefore if the incident radiation is represented by the complex quantity $R\exp(j\omega t)$ the contribution in the direction θ due to dx is given by $Rf(x)\exp j(\omega t - Kx)$. Hence the total contribution from the whole screen is given by

$$\int_{-\infty}^{\infty} Rf(x)\exp j(\omega t - Kx)\,dx = R\exp(j\omega t)\int_{-\infty}^{\infty} f(x)\exp(-jKx)\,dx$$

and the radiation in the direction θ, relative to the incident radiation may be written

$$\int_{-\infty}^{\infty} f(x)\exp(-jKx)\,dx = F(K), \quad K = 2\pi \sin \theta/\lambda$$

Hence we see that $f(x)$ and $F(K)$ form a Fourier pair. Note that $Kx = (2\pi \sin \theta/\lambda)x$ is dimensionless as required. The function $f(x)$ represents the transmission characteristic of the screen. The complex function $F(K)$ represents, in both magnitude and phase, the vector defining the radiation in the direction θ (where $2\pi \sin \theta/\lambda = K$). In this analysis we have assumed that no phase change is introduced by the screen and so $f(x)$ is a real function. The function $F(K)$ however does contain a phase term and so is complex. This case then corresponds to the most common case arising with the pair $f(t)$ and $F(\omega)$ in which $f(t)$ is real and $F(\omega)$ in general complex.

6.3 Diffraction at a Slit

Consider the case of diffraction at a slit. Let the slit extend from $x = -\delta$ to $x = +\delta$ as shown in Fig. 6.2(a). Suppose the amplitude of

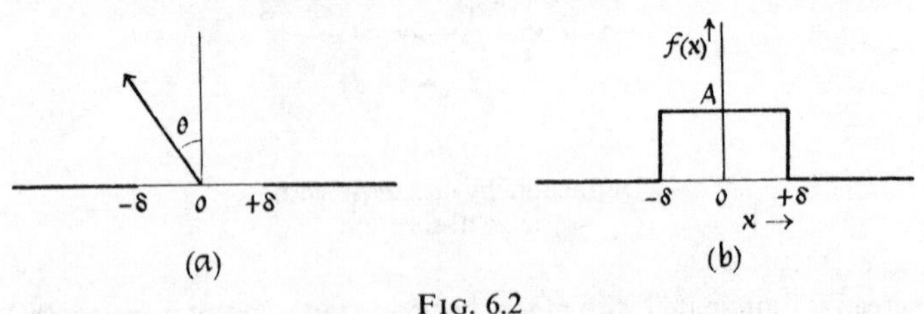

FIG. 6.2
Diffraction by a single slit

the light transmitted by the slit is A and that the screen is completely opaque elsewhere. Then $f(x)$ is as shown in Fig. 6.2(*b*). If we let $A\delta = 1$ then from the results of section 4.2 we can write down immediately

$$F(K) = \frac{\sin K\delta}{K\delta} = \frac{\sin(2\pi\delta\sin\theta/\lambda)}{(2\pi\delta\sin\theta/\lambda)}$$

The intensity of the radiation is proportional to the square of the amplitude and hence varies like

$$I = \frac{\sin^2(2\pi\delta\sin\theta/\lambda)}{(2\pi\delta\sin\theta/\lambda)^2}$$

which is a well-known formula in physical optics.

6.4 The Diffraction Grating

The formula for the diffraction grating can be developed in a similar manner to that for the slit. The case of the slit corresponds to a pulse and the grating corresponds to a finite train of pulses. Hence we have immediately from section 4.6

$$I = |F(K)|^2 = \left(\frac{\sin K\delta}{K\delta}\frac{\sin NKd}{\sin Kd}\right)^2$$

as the variation in intensity produced by a diffraction grating consisting of N slits of width 2δ and spacing d. This again is a well-known formula in physical optics.

It is of interest to note the effect of displacing the grating in the x direction. If the amount of the displacement is x_0 then $f(x-x_0)$ represents the variation of the transmission characteristic with x. By the shift theorem this gives an amplitude variation $\exp(-jKx_0)F(K)$, and the intensity is given by

$$|\exp(-jKx_0)\,F(K)|^2 = |F(K)|^2$$

Thus the variation of intensity is not affected by displacement of the grating.

We now consider the case of a sinusoidal grating. In this case the transmission characteristic varies sinusoidally across the grating and we have

$$f(x) = 1 + a\cos(2\pi x/l)$$

By the superposition theorem this will give rise to a transform consisting of the sum of the transform of 1 and of the transform of $a\cos(2\pi x/l)$. The first term corresponds to a beam in the direction $\theta = 0$, i.e. an undeflected beam and the second gives rise to two components in the directions given by $K = \pm 2\pi/l$, or

$$2\pi \sin\theta/\lambda = \pm 2\pi/l$$

Hence $\sin\theta = \pm\lambda/l$.

The beams in this case are infinitely narrow since we assumed a grating of infinite extent. In optical terminology the whole of the energy of the incident beam becomes concentrated in the zero and first-order spectra after transmission through a sinusoidal grating.

6.5 Amplitude Modulation

Suppose we have a waveform $h(t)$ such that $|h(t)| \leqslant 1$ for all t and let $H(\omega)$ be the Fourier transform of $h(t)$. If now we modulate the amplitude of a carrier $\cos\omega_c t$ by $h(t)$ then we have the resulting waveform $f(t)$

$$f(t) = \{1+h(t)\}\cos\omega_c t$$

Expanding the function we have

$$f(t) = \cos\omega_c t + \tfrac{1}{2}\exp(j\omega_c t)\,h(t) + \tfrac{1}{2}\exp(-j\omega_c t)\,h(t)$$

Therefore the spectrum of the modulated carrier consists of two spectral lines representing the carrier plus, by the shift theorem, the

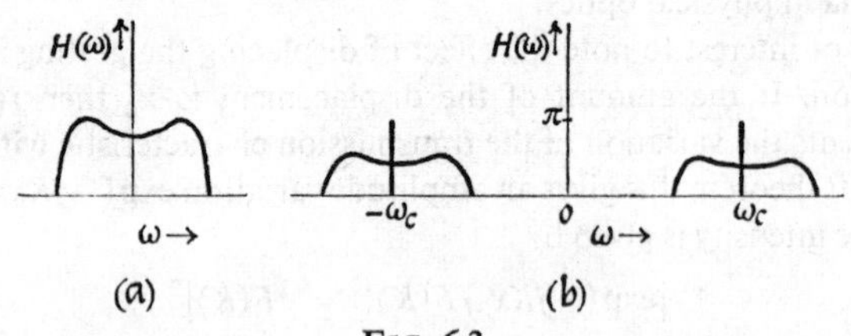

FIG. 6.3
The development of the spectrum for amplitude modulation

two terms $\tfrac{1}{2}H(\omega+\omega_c)$ and $\tfrac{1}{2}H(\omega-\omega_c)$. Suppose for simplicity $H(\omega)$ is real, then as we have seen, it must be symmetrical about $\omega = 0$ as shown in Fig. 6.3(*a*). The spectrum of the modulated wave is then obtained by halving the spectrum and translating one half to the point $\omega = \omega_c$ and the other to $\omega = -\omega_c$. Note that from this point of view negative frequencies of $H(\omega)$ after translation to the new origin will appear in the region of positive frequencies in the spectrum for the modulated wave. To complete the spectrum two lines must be added to represent the carrier.

Suppose for example $h(t) = \cos pt$. Then the spectrum is given by two impulses (or lines) of amplitude π (section 4.5) at frequencies $\omega = \pm p$. The spectrum of a carrier when modulated by this function can be obtained by the construction just described. The result is shown in Fig. 6.4(*b*) and the two familiar sidebands removed from

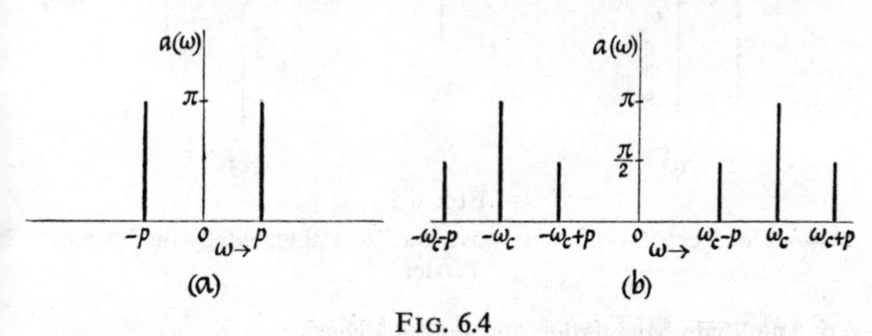

FIG. 6.4

(*a*) The spectrum of a cosine wave and (*b*) a cosinusoidally modulated carrier

the carrier by $\pm p$ appear. The spectrum $F(\omega)$ of $f(t)$ in this case is real and Fig. 6.4 can be considered as a plot of $\text{Re}\{F(\omega)\}$ or of $a(\omega)$ as defined in Chapter III.

Now consider the case of sinusoidal modulation of a cosinusoidal carrier, i.e.

$$h(t) = \sin pt$$

and
$$f(t) = (1 + \sin pt) \cos \omega_c t$$

The complex spectrum $H(\omega)$ is now pure imaginary and

$$\text{Im}\{H(\omega)\} = -jb(\omega)$$

(section 3.4). We also know that the spectrum must be anti-symmetrical. If we plot $b(\omega)$ then we get two spectral lines, one of magnitude $+\pi$ at frequency $+p$ and one of magnitude $-\pi$ at $-p$, as shown in Fig. 6.5(*a*). Applying the construction developed above gives a spectrum shown in Fig. 6.5(*b*). Note that the construction leads to a spectrum which is also anti-symmetric as $b(\omega)$ must always be. The complete spectrum for $f(t)$ also contains a real part $a(\omega)$ consisting of two spectral lines of magnitude π and frequencies $\pm\omega_c$ representing the carrier.

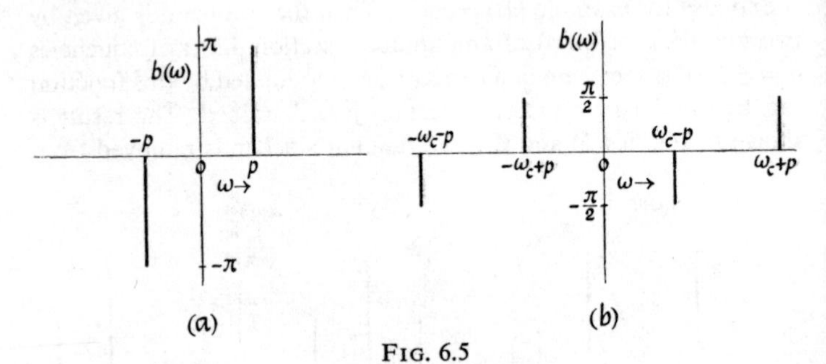

FIG. 6.5

(*a*) The spectrum of a sine wave and (*b*) a sinusoidally modulated carrier

6.6 Amplitude Modulation and Paired Echoes

In the last section we considered a time function which was modulated in amplitude sinusoidally. In section 5.8 we considered a function of frequency whose magnitude varied sinusoidally. We shall now demonstrate how these two effects are related by the principle of duality.

Consider first the case of amplitude modulation of a sinusoidal carrier $\sin\omega_c t$. The modulated wave is represented by

$$f(t) = \{1+h(t)\}\sin\omega_c t$$

$$= \{1+h(t)\}\frac{\exp(j\omega_c t)-\exp(-j\omega_c t)}{2j}$$

$$\therefore \{1+h(t)\}j\sin\omega_c t = \tfrac{1}{2}\{1+h(t)\}\{\exp(j\omega_c t)-\exp(-j\omega_c t)\}$$

If $h(t) = \cos pt$ then the spectrum $H(\omega)$ is real and hence the spectrum of $(1+\cos pt)j\sin\omega_c t$ is also seen to be real. The spectrum can be obtained from the spectrum of $\cos pt$ and the shift theorem and is shown plotted in Fig. 6.6(*b*). It may be noted that although this spectrum is real it is not symmetrical. The reason for this is that the symmetry of $a(\omega)$ was deduced on the assumption that the original function of time was pure real and in this case this is not so. Figure

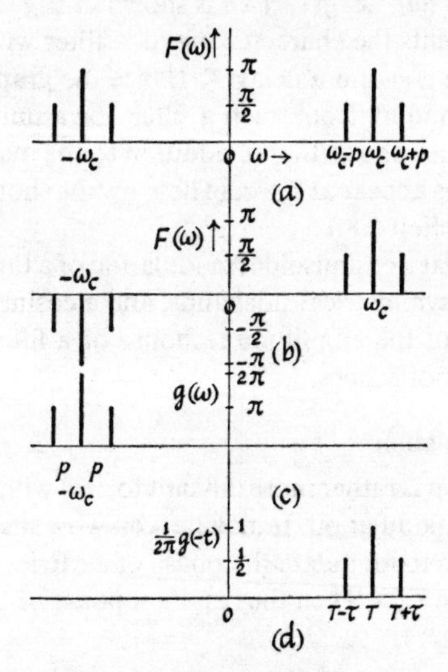

FIG. 6.6

Development of response of filter with imperfect amplitude characteristic from amplitude modulation spectrum using the duality theorem

6.6(*a*) shows the spectrum for the function $(1+\cos pt)\cos\omega_c t$, and the spectrum for the function

$$f(t) = (1+\cos pt)(\cos\omega_c t - j\sin\omega_c t) = (1+\cos pt)\exp(-j\omega_c t)$$

is obtained by the superposition theorem and is shown in Fig. 6.6(*c*). Let this function be $g(\omega)$. Then the signal of $g(\omega)$ is $f(t)$, therefore by the principle of duality the signal of $f(\omega)$ is $g(-t)/2\pi$. The expression for $f(\omega)$ is obtained from the expression for $f(t)$ by replacing the variable t by the variable ω and replacing the particular values of frequency p and ω_c by the particular values of time τ and T. Hence

$$f(\omega) = (1+\cos\omega\tau)\exp(-j\omega T)$$

The signal of $f(\omega)$, i.e. $g(-t)/2\pi$ is shown in Fig. 6.6(d). The function $f(\omega)$ represents the characteristic of a filter with an amplitude response $(1+\cos\omega\tau)$ and a delay T. Hence the graph of Fig. 6.6(d) represents the output from such a filter for a unit impulse input (section 5.7), which shows that in addition to the main signal at time T a pair of echoes appear at $T\pm\tau$. (These results should be compared with those of section 5.8.)

We see then that a cosinusoidal modulation of a time function gives rise to a pair of symmetrical sidebands, and a cosinusoidal variation with frequency of the amplitude response of a filter gives rise to a symmetrical pair of echoes.

6.7 Phase Modulation

Phase modulation is rather more difficult to deal with mathematically than amplitude modulation. It may be seen why this is so as follows. Suppose we wish to modulate the phase of a carrier of frequency ω_c with the function $m(t)$. Then the resultant phase of the wave is given by

$$\Phi(t) = \omega_c t+m(t)$$

and the signal is

$$\begin{aligned} f(t) &= \cos\{\omega_c t+m(t)\} \\ &= \tfrac{1}{2}[\exp(j\omega_c t)\exp\{jm(t)\}+\exp(-j\omega_c t)\exp\{-jm(t)\}] \end{aligned}$$

The function $m(t)$ now appears within the exponent and this makes the general case difficult to deal with. We shall therefore consider the case of sinusoidal modulation and for simplicity construct the case for $f(t)$ symmetric so that the spectrum is pure real. Let

$$\begin{aligned} f(t) &= \phi\{\omega_c t+m(t)\} \\ &= \phi(\tau) \end{aligned}$$

where $$\tau(t) = \omega_c t+m(t)$$

We wish to construct the case for $f(t)$ symmetric, i.e. $f(-t) = f(t)$. If $f(t)$ is symmetric then $\phi(\tau)$ is also symmetric

$$\therefore \phi(\tau) = \phi(-\tau)$$

Also, since $\quad f(t) = \phi(\tau), \quad f(-t) = \phi(-\tau)$

Now $\quad f(-t) = \phi\{-\omega_c t + m(-t)\}$

and $\quad \phi(-\tau) = \phi\{-\omega_c t - m(t)\}$

hence $m(-t) = -m(t)$ and $m(t)$ must be anti-symmetric. Thus since we wish to consider sinusoidal modulation we consider

$$m(t) = \phi_0 \sin pt$$

The resulting modulated wave is then given by

$$f(t) = \cos\{\omega_c t + \phi_0 \sin pt\}$$
$$= \tfrac{1}{2}[\exp(j\omega_c t)\exp(j\phi_0 \sin pt) + \exp(-j\omega_c t)\exp(-j\phi_0 \sin pt)]$$

Now from the theory of Bessel functions we have

$$\exp(j\phi_0 \sin pt) = \sum_{n=-\infty}^{\infty} J_n(\phi_0)\exp(jnpt)$$

$$\therefore \exp(j\phi_0 \sin pt) = J_0(\phi_0) + J_1(\phi_0)\exp(jpt) + J_{-1}(\phi_0)\exp(-jpt) + \ldots$$

We can expand the expression $\exp(-j\phi_0 \sin pt)$ by noting that $\exp\{-j\phi_0 \sin(pt)\} = \exp\{j\phi_0 \sin(-pt)\}$

$$\therefore \exp\{-j\phi_0 \sin pt\} = J_0(\phi_0) + J_1(\phi_0)\exp(-jpt) + J_{-1}(\phi_0)\exp(jpt) + \ldots$$

Hence expanding the expression for $f(t)$ we have

$$f(t) = \tfrac{1}{2}[J_0(\phi_0)\exp(j\omega_c t) + J_1(\phi_0)\exp\{j(\omega_c + p)t\} + J_{-1}(\phi_0)\exp\{j(\omega_c - p)t\} + \ldots + J_0(\phi_0)\exp(-j\omega_c t) + J_1(\phi_0)\exp\{-j(\omega_c + p)t\} + J_{-1}(\phi_0)\exp\{-j(\omega_c - p)t\} + \ldots]$$

$$\therefore f(t) = J_0(\phi_0)\cos\omega_c t + J_1(\phi_0)\cos(\omega_c + p)t + J_{-1}(\phi_0)\cos(\omega_c - p)t + J_2(\phi_0)\cos(\omega_c + 2p)t + J_{-2}(\phi_0)\cos(\omega_c - 2p)t + \ldots$$

This expression may also be written

$$f(t) = \sum_{n=-\infty}^{\infty} J_n(\phi_0)\cos(\omega_c+np)t$$

Thus we see that for sinusoidal modulation we get an infinite series of spectra spaced at multiples of the modulation frequency above and below the carrier. If there were more than one modulating frequency

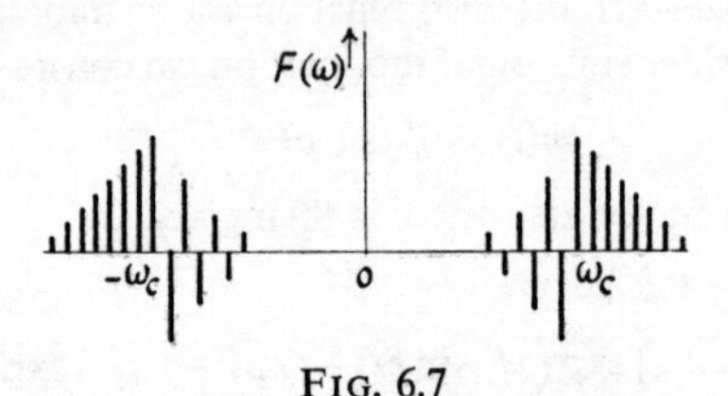

FIG. 6.7
The spectrum of a phase modulated carrier (diagrammatic)

then it may be shown that in addition to the two infinite sets of sidebands corresponding to the two frequencies, all the beat frequencies also occur. This is why the general case is so complex.

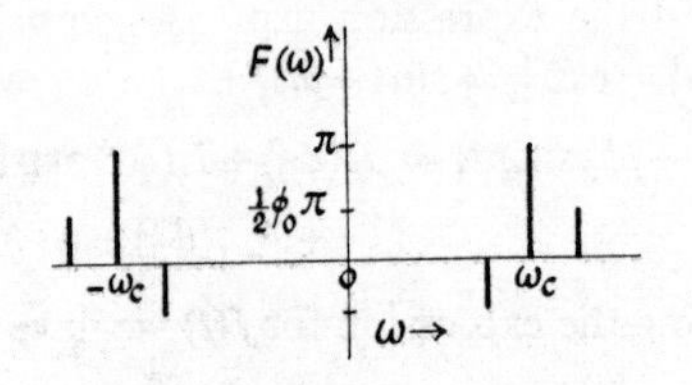

FIG. 6.8
The spectrum for small phase deviation

If the maximum phase excursion ϕ_0 is small then we can sketch the spectral lines bearing in mind that $J_{-n}(x) = (-1)^n J_n(x)$. The sketch is shown in Fig. 6.7. The way in which the spectrum develops as ϕ_0 becomes larger is discussed in works on phase modulation and frequency modulation (see, for example, Arguimbau and Stuart (1956)). If ϕ_0 is sufficiently small then $J_0(\phi_0) = 1$ and $J_1(\phi_0) = \frac{1}{2}\phi_0$ and we may neglect all higher order terms. The expression for $f(t)$ then becomes

$$f(t) = \cos \omega_c t + \tfrac{1}{2}\phi_0 \cos(\omega_c + p)\, t - \tfrac{1}{2}\phi_0 \cos(\omega_c - p)\, t$$

This expression corresponds closely to that for amplitude modulation, the difference being that the lower sideband is negative. The spectrum for this case is shown in Fig. 6.8.

6.8 Phase Modulation and Paired Echoes

In the same way as we related amplitude modulation to distortion of a filter amplitude characteristic so we can relate phase modulation to distortion of a filter phase characteristic by the principle of duality.

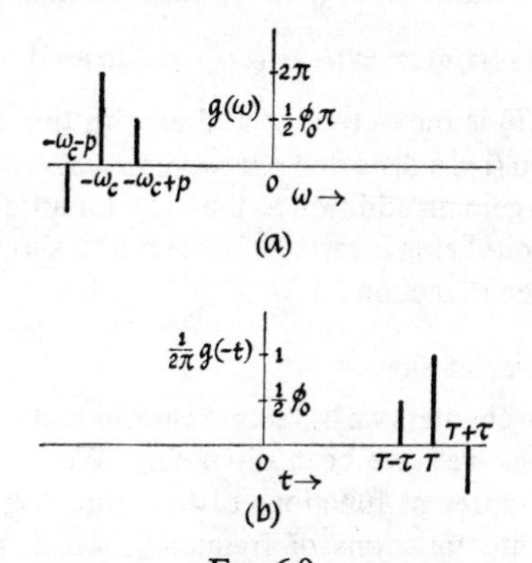

FIG. 6.9

Development of response of filter with imperfect phase characteristic from phase modulation spectrum using the duality theorem

Consider the case just discussed in which ϕ_0 is small so that only terms of the first order need be taken into account. Following the method used for the amplitude case we form a spectrum in the negative region of ω. Thus we consider

$$\begin{aligned} f(t) &= \exp\{-j\omega_c t + j\phi_0 \sin pt\} \\ &= J_0(\phi_0)\exp(-j\omega_c t) + J_1(\phi_0)\exp(-j\omega_c t + jpt) \\ &\quad + J_{-1}(\phi_0)\exp(-j\omega_c t - jpt) \\ &= \exp(-j\omega_c t) + \tfrac{1}{2}\phi_0 \exp\{-j(\omega_c - p)t\} \\ &\quad - \tfrac{1}{2}\phi_0 \exp\{-j(\omega_c + p)t\} \end{aligned}$$

Hence the spectrum $g(\omega)$ is as shown in Fig. 6.9(*a*). By the principle of duality if $f(t)$ is the signal of $g(\omega)$ then the signal of $f(\omega)$ is

$$g(-t)/2\pi$$

Hence the signal shown in Fig. 6.9(*b*) has spectrum

$$f(\omega) = \exp\{-j(\omega T - \phi_0 \sin \omega\tau)\}$$

Hence Fig. 6.9(*b*) is the output of a filter with this characteristic as transfer function (i.e. a distorted phase characteristic) when the input is an impulse. Again, in addition to the main impulse at T two echoes occur, the later one being inverted. (These results should be compared to those obtained in section 5.8.)

6.9 Vector Representation

In the preceding chapters we have used various methods of representing the functions we have been discussing. We have used analytic expressions to represent functions of time directly; we have transformed these into functions of frequency, which in general were complex, using either an analytic expression or alternatively plotting the real and imaginary parts of the spectrum graphically; and we have used the idea of vectors (which vary with time) plotted on an Argand diagram. In this section we shall consider the vector representation of modulated waves.

We saw in section 3.6 that a cosine function can be considered as the sum of two rotating vectors representing the functions $\frac{1}{2}\exp(j\omega_c t)$ and $\frac{1}{2}\exp(-j\omega_c t)$. These vectors produce a real variation $\cos w_c t$.

The relative phase of the vectors is obtained most readily by considering $t = 0$ and at this instant they both lie along the positive real axis. The spectrum is pure real and consists of two lines at $\pm\omega_c$ of

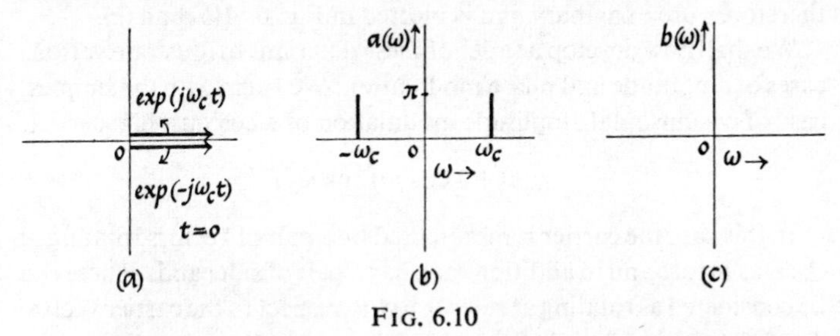

FIG. 6.10
(*a*) Argand diagram for vector representation and (*b*) and (*c*) the spectrum of a cosine wave

magnitude π. This can be summarized diagrammatically as shown in Fig. 6.10. Figure 6.10(*a*) is the Argand diagram drawn for $t = 0$. The two vectors should be coincident but they are drawn slightly separated

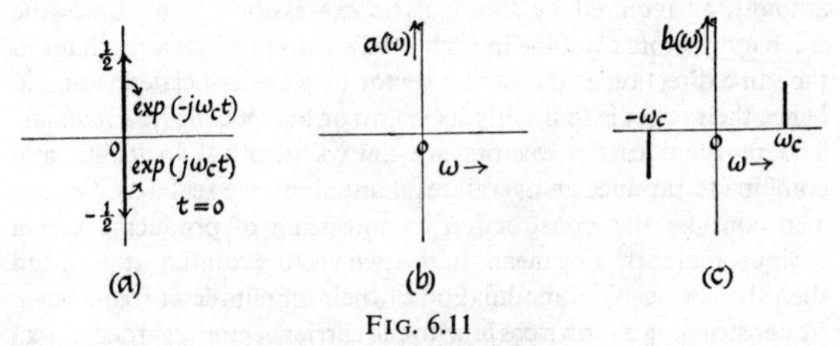

FIG. 6.11
(*a*) Argand diagram for vector representation and (*b*) and (*c*) the spectrum of a sine wave

for clarity. Figures 6.10(*b*) and 6.10(*c*) show the plots of $a(\omega)$ and $b(\omega)$, the latter being zero everywhere in this case.

The case for $\sin\omega_c t$ is shown in Fig. 6.11. In order to produce a real variation of $\sin\omega_c t$, i.e. a variation along the x axis, we multiply the positively rotating vector by $-j$ so that at time $t = 0$ it lies along the negative y axis and the negatively rotating vector by $+j$. At time $t = 0$

the two vectors clearly cancel and the resultant is zero and in general we have $\frac{1}{2}j\{\exp(-j\omega_c t) - \exp(j\omega_c t)\}$ or $\{\exp(j\omega_c t) - \exp(-j\omega_c t)\}/2j$, which gives a real variation $\sin\omega_c t$ as required. The spectrum is therefore pure imaginary and is plotted in Fig. 6.11(*b*) and (*c*).

We shall now develop a series of these diagrams to illustrate various cases of amplitude and phase modulation. We begin with the simplest case of cosinusoidal amplitude modulation of a cosinusoidal carrier.

$$f(t) = (1 + a\cos pt)\cos\omega_c t$$

In this case the carrier is represented by a pair of vectors rotating at $\pm\omega_c$ as before and in addition each has a pair of sidebands. These can be considered as rotating at rates $\pm p$ with respect to the carrier vector. Thus the sideband vectors can be drawn from the tip of the carrier vector and must then be considered to be carried round with it. At $t = 0$ the situation is as shown in the Figure. (*Note:* The vectors are actually coincident with the real axis but are shown separated for clarity.) The sidebands at this instant combine in such a way as to increase the magnitude of the resultant vector by the maximum amount, as required by the analytic expression. At all times the sideband vectors combine in such a way as to produce a resultant in the same direction as the carrier vector they are associated with and hence their effect is to modify its length or to *modulate its amplitude*. The resultant carrier vectors are always identical in length and combine to produce an overall resultant along the real axis. We can also consider this construction as consisting of producing first a cosinusoidal carrier by means of the two vectors rotating at $\pm\omega_c$ and then the cosinusoidal modulation of their amplitude at frequency p by constructing a term $a\cos pt$ with the carrier vector regarded as real axis. Thus at $t = 0$ the vectors rotating at $\pm p$ lie along the carrier vectors, for this case along the real axis of the Argand diagram. Since at $t = 0$ all vectors lie along the positive real axis the spectrum is pure real and all terms are positive. The plots of $a(\omega)$ and $b(\omega)$ are shown in Fig. 6.12.

The second case is when the modulation is sinusoidal and the resulting analytic function becomes

$$f(t) = (1 + a\sin pt)\cos\omega_c t$$

This case is shown in Fig. 6.13. The same considerations apply in this case. The vectors representing the sidebands are constructed at the tip of the carrier vectors and at $t = 0$ combine in such a way as to leave the length of the carrier vector unmodified. Again the sideband vectors always combine to produce a resultant in the same direction as their associated carrier and the two carriers combine to produce an overall resultant along the real axis. Alternatively, we can regard this construction as before as first producing the carrier by means of two vectors rotating at speeds $\pm\omega_c$ and then producing sinusoidal modulation of their amplitude by constructing the $a \sin pt$ term with the carrier vector regarded as real axis. Thus at $t = 0$ the sideband vectors all lie parallel to the imaginary axis in the Argand diagram. Hence the complete complex spectrum has a real part consisting of two lines at $\pm\omega_c$ and an imaginary part consisting of four lines at $(\omega_c \pm p)$ and $(-\omega_c \pm p)$. These are shown in Fig. 6.13(*b*) and (*c*). Note that the sideband vector in Fig. 6.13(*a*) associated with the $+\omega_c$ vector and rotating positively, i.e. corresponding to the sideband $(\omega_c + p)$ is drawn in the negative direction and hence corresponds to a positive line in $b(\omega)$ since $b(\omega) = -\text{Im}\{F(\omega)\}$. We can also obtain the spectrum from the construction developed in section 6.5. The modulating waveform is $\sin pt$ and its spectrum consists of two lines as shown in Fig. 6.11(*c*). We halve the magnitude of these lines and translate the result to the new origin which gives the same result as before, shown in Fig. 6.13(*c*). To complete the spectrum the real part consisting of two lines representing $\cos \omega_c t$ must be added.

Two more cases arise for amplitude modulation, when the carrier sinusoidal. The analytic expressions are $(1 + a \cos pt) \sin \omega_c t$ and $(1 + a \sin pt) \sin \omega_c t$. These may be treated in a similar way and the results are shown in Figs. 6.14 and 6.15.

We can consider the four similar cases for phase modulation. For simplicity we shall suppose that the maximum phase modulation ϕ_0 is small so that only the first pair of sidebands need be considered (see section 6.7). Considering first the case $f(t) = \cos(\omega_c t + \phi_0 \sin pt)$ we must produce a variation at right angles to the carrier vector in order to modulate its phase. Hence the sideband vectors are as shown in Fig. 6.16(*a*) and it is seen that at $t = 0$ all vectors lie along the real axis and the spectrum is therefore pure real. This shows

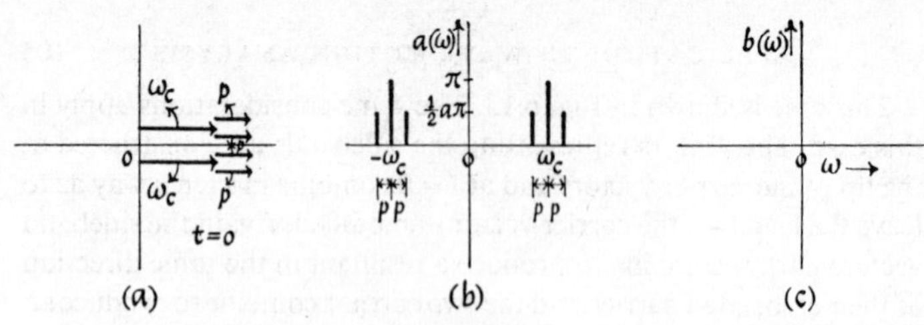

FIG. 6.12

The function $(1+a\cos pt)\cos\omega_c t$

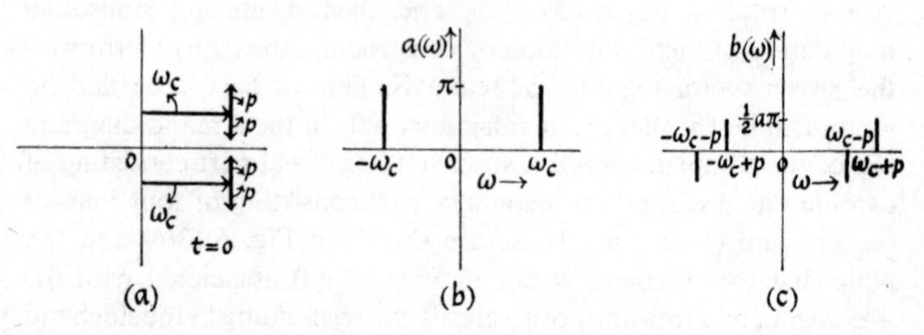

FIG. 6.13

The function $(1+a\sin pt)\cos\omega_c t$

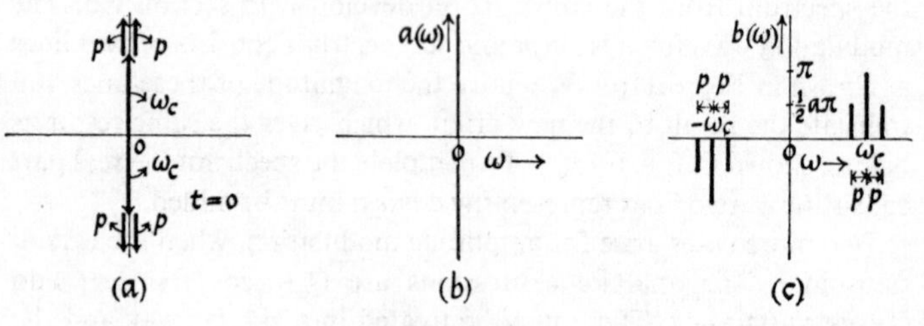

FIG. 6.14

The function $(1+a\cos pt)\sin\omega_c t$

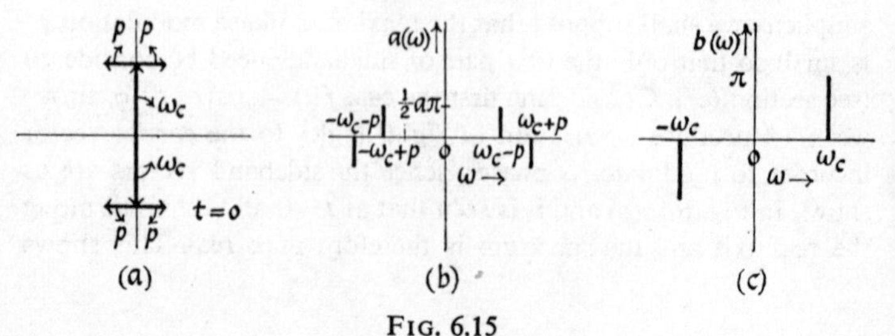

FIG. 6.15

The function $(1+a\sin pt)\sin\omega_c t$

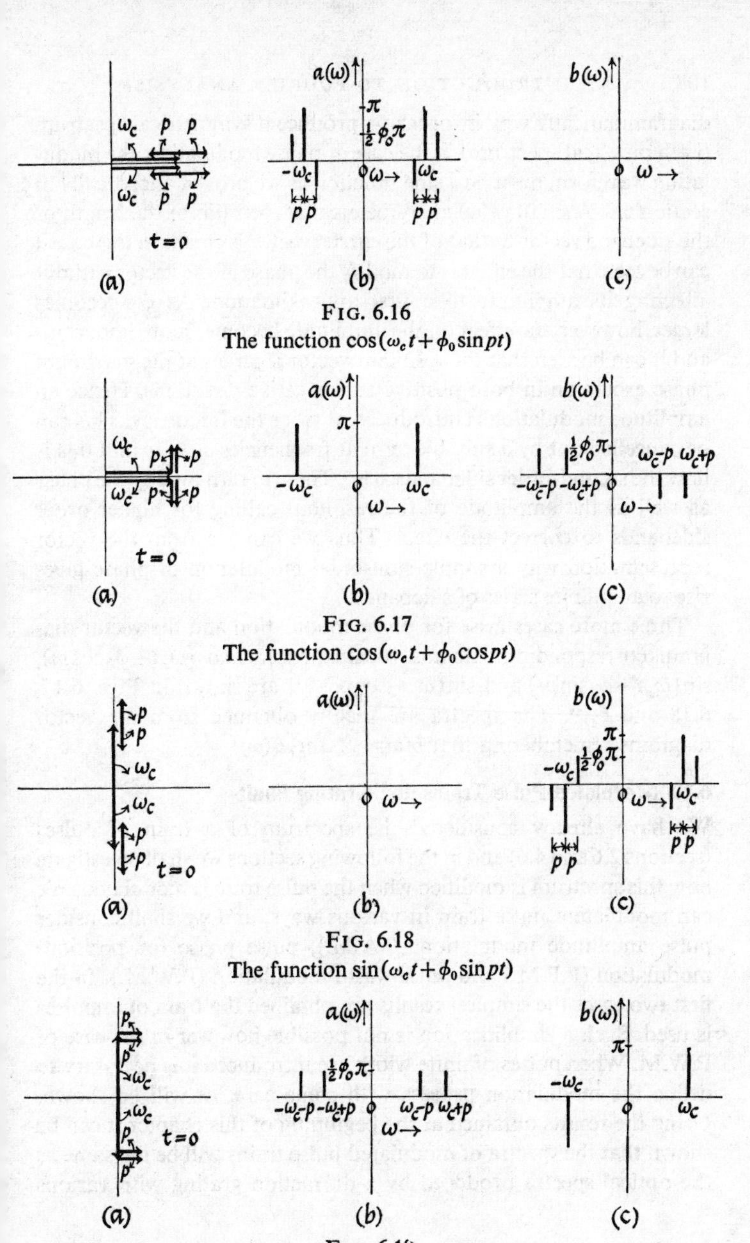

FIG. 6.16
The function $\cos(\omega_c t + \phi_0 \sin pt)$

FIG. 6.17
The function $\cos(\omega_c t + \phi_0 \cos pt)$

FIG. 6.18
The function $\sin(\omega_c t + \phi_0 \sin pt)$

FIG. 6.19
The function $\sin(\omega_c t + \phi_0 \cos pt)$

diagrammatically why in order to produce a symmetrical spectrum (i.e. a pure real spectrum) in the case of phase modulation, the modulating waveform must be a sine function as we proved analytically in section 6.7. A small value of ϕ_0 means that the ratio of the length of the sideband vector to that of the carrier vector is small. In this case it can be seen that the effect is to modify the phase of the vector without affecting its amplitude to a first approximation. As ϕ_0 becomes larger, however, the effect on the amplitude becomes more important and it can be seen that the resultant vector is larger at the maximum phase excursion in both positive and negative directions. Hence an amplitude modulation is introduced at twice the frequency. This can be cancelled out by a suitable term at frequencies $\omega_c \pm 2p$ and this is how the second-order sidebands arise. These in turn modify the phase as well as the amplitude of the resultant calling for higher order sidebands to correct the effect. Thus we can see from the vector representation why a simple sinusoidal modulation of phase gives rise to an infinite series of sidebands.

Three more cases arise for phase modulation and the vector diagrams corresponding to the analytical expressions $\cos(\omega_c t + \phi_0 \cos pt)$, $\sin(\omega_c t + \phi_0 \sin pt)$ and $\sin(\omega_c t + \phi_0 \cos pt)$ are shown in Figs. 6.17, 6.18 and 6.19. The spectra are readily obtained from the vector diagrams remembering that $b(\omega) = -\text{Im}\{F(\omega)\}$.

6.10 Modulated Pulse Trains and Grating Faults

We have already considered the spectrum of a train of pulses (sections 2.6 and 4.6) and in the following sections we shall investigate how this spectrum is modified when the pulse train is modulated. We can modulate a pulse train in various ways, and we shall consider pulse amplitude modulation (P.A.M.), pulse phase (or position) modulation (P.P.M.) and pulse width modulation (P.W.M.). In the first two cases the simplest results are obtained if a train of impulses is used. Such a simplification is not possible however in the case of P.W.M. When pulses of finite width are introduced it is necessary to define the modulation process with some care, as will be shown. Using the results obtained at the beginning of this chapter it can be shown that the spectra of modulated pulse trains will be the same as the optical spectra produced by a diffraction grating with various types of periodic faults in the ruling. The two sets of phenomena will be related in subsequent sections.

6.11 Pulse Amplitude Modulation

We shall consider a train of impulses modulated in amplitude by a cosine function as shown in Fig. 6.20(a). Suppose the unmodulated

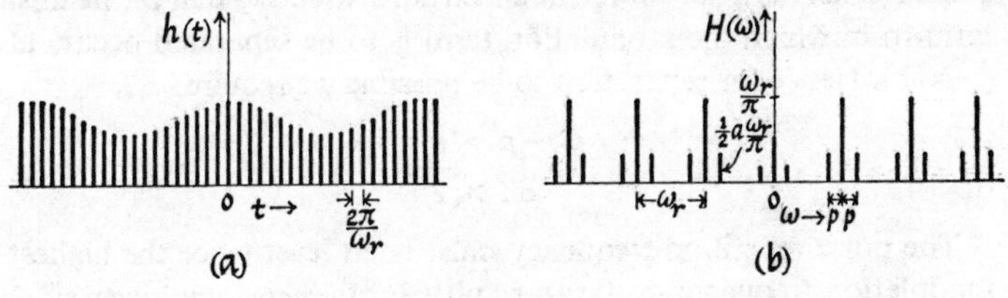

FIG. 6.20

(a) The pulse amplitude modulation signal and (b) its spectrum

pulse train is $f(t)$ and has spectrum $F(\omega)$. Then the modulated train is given by

$$\begin{aligned} h(t) &= (1+a\cos pt)f(t) \\ &= f(t)+\tfrac{1}{2}a\exp(jpt)f(t)+\tfrac{1}{2}a\exp(-jpt)f(t) \end{aligned}$$

Hence by the shift theorem the spectrum of the modulated train is given by

$$H(\omega) = F(\omega)+\tfrac{1}{2}a\,F(\omega-p)+\tfrac{1}{2}a\,F(\omega+p)$$

The resulting spectrum is shown in Fig. 6.20(b). The value of $\omega_r/2\pi$ for the magnitude of the spectra is that appropriate to a train of unit impulses occurring at intervals $2\pi/\omega_r$. In this case of course the height of the pulses would be infinite. If on the other hand we consider impulses of finite height then the magnitude of the spectral lines is zero as we saw in section 2.6. To produce a case in which both the time and frequency functions are finite we must consider pulses of finite width. In this case the spectral lines are modified in amplitude by the term $\sin(k\omega_r\tau)/(k\tau)$ where k is the order of the spectrum and 2τ is the pulse width. The function $h(t)$ was obtained by multiplying $f(t)$ by the factor $(1+a\cos pt)$ so that at each instant of time $f(t)$ is modified by the value of $(1+a\cos pt)$ at that same instant. Thus the resulting pulses are not rectangular pulses and the modulation modifies their shape as well as their amplitude.

From the spectrum shown in Fig. 6.20(*b*) we see that a low pass filter can be used to separate the term at $\omega = p$ from the rest of the terms and hence produce the original modulating wave. Thus a low pass filter acts as a detector. The spectrum also shows that the nearest term from which the modulation term is to be separated occurs at $(\omega_r - p)$. Hence for separation to be possible we require

$$\omega_r - p > p$$

or

$$\omega_r > 2p$$

The pulse repetition frequency must be at least twice the highest modulation frequency to be transmitted, otherwise the lower side-bands of the first harmonic of the repetition frequency overlap with terms occurring at the modulation frequencies and it becomes impossible to recover the original modulating waveform without distortion.

6.12 Ghost Spectra

The function h of the last section could represent a diffraction grating in which the absorption of the transparent portion had a sinusoidal variation with distance. Such an effect is produced by a variation in the heaviness of the ruling of the lines. The function H would then represent the resulting spectrum produced, as we saw in section 6.2. Each spectral line is accompanied by two fainter lines equally spaced on each side of it. These have been termed ghosts. The intensity of the ghost produced by this type of fault is independent of the order of the spectral line it accompanies as can be seen from the plot of H in Fig. 6.20(*b*).

6.13 Pulse Phase Modulation

We shall first consider the simpler case of a train of impulses. We may write the expression for these

$$f(t) = \frac{\omega_r}{2\pi} \sum_{k=-\infty}^{\infty} \exp(jk\omega_r t)$$

We now, by analogy with the method used in the derivation of phase modulation, replace $\omega_r t$ by $(\omega_r t+\phi_0 \sin pt)$. Hence we have

$$h(t) = \frac{\omega_r}{2\pi}\sum_{k=-\infty}^{\infty} \exp\{jk(\omega_r t+\phi_0 \sin pt)\}$$

$$= \frac{\omega_r}{2\pi}\sum_{k=-\infty}^{\infty} \exp(jk\omega_r t)\exp(jk\phi_0 \sin pt)$$

and using the relation from Bessel function theory

$$\exp(jk\phi_0 \sin pt) = \sum_{n=-\infty}^{\infty} J_n(k\phi_0)\exp(jnpt)$$

we have

$$f(t) = \frac{\omega_r}{2\pi}\sum_{k=-\infty}^{\infty} \exp(jk\omega_r t)\sum_{n=-\infty}^{\infty} J_n(k\phi_0)\exp(jnpt)$$

$$= \frac{\omega_r}{2\pi}\sum_{k=-\infty}^{\infty}\sum_{n=-\infty}^{\infty} J_n(k\phi_0)\exp\{j(k\omega_r+np)t\}$$

We can simplify this as follows. If $k=0$ then $J_n(k\phi_0)=J_n(0)=0$, $n\neq 0$. Hence for $k=0$ the second summation consists of only one term, viz. $J_0(0)\exp(0)=1$. We can therefore write

$$f(t) = \frac{\omega_r}{2\pi}\left[1+\sum_{n=-\infty}^{\infty}\left(\sum_{k=1}^{\infty}+\sum_{k=-1}^{-\infty}\right)J_n(k\phi_0)\exp\{j(k\omega_r+np)t\}\right]$$

$$= \frac{\omega_r}{2\pi}\left[1+\sum_{n=-\infty}^{\infty}\sum_{k=1}^{\infty}\{J_n(k\phi_0)\exp j(k\omega_r+np)t\right.$$

$$\left.+J_n(-k\phi_0)\exp j(-k\omega_r+np)t\}\right]$$

$$= \frac{\omega_r}{2\pi}\left[1+\sum_{n=-\infty}^{\infty}\sum_{k=1}^{\infty}\{J_n(k\phi_0)\exp j(k\omega_r+np)t\right.$$

$$\left.+J_{-n}(-k\phi_0)\exp[-j(k\omega_r+np)t]\}\right]$$

and since

$$J_{-n}(-x) = J_n(x)$$

$$f(t) = \frac{\omega_r}{2\pi}\left[1 + \sum_{n=-\infty}^{\infty} \sum_{k=1}^{\infty} J_n(k\phi_0)\{\exp j(k\omega_r + np)t + \exp[-j(k\omega_r + np)t]\}\right]$$

$$\therefore f(t) = \frac{\omega_r}{2\pi}\left[1 + 2\sum_{n=-\infty}^{\infty} \sum_{k=1}^{\infty} J_n(k\phi_0)\cos(k\omega_r + np)t\right]$$

Expanding the summation terms we have

$$f(t) = \frac{\omega_r}{\pi}[\tfrac{1}{2} + J_0(\phi_0)\cos\omega_r t + J_1(\phi_0)\cos(\omega_r + p)t$$
$$+ J_{-1}(\phi_0)\cos(\omega_r - p)t + \ldots$$
$$+ J_0(2\phi_0)\cos 2\omega_r t + J_1(2\phi_0)\cos(2\omega_r + p)t$$
$$+ J_{-1}(2\phi_0)\cos(2\omega_r - p)t + \ldots + \ldots]$$

Figure 6.21 is a diagrammatic sketch of the spectra for small deviation ϕ_0. It consists of a series of lines at the harmonics of the repetition frequency each having an infinite set of sidebands similar

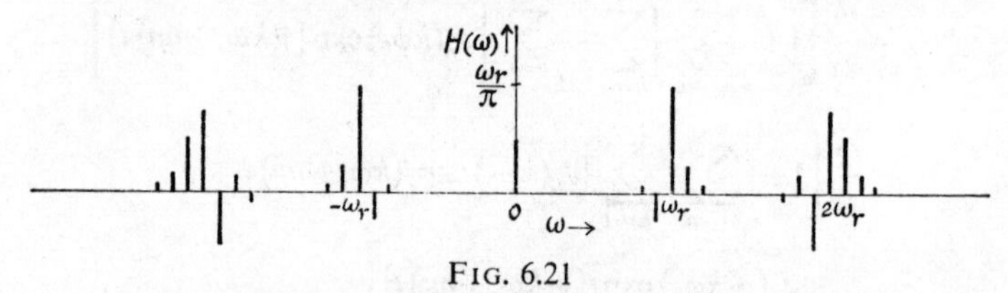

FIG. 6.21

The pulse phase modulation spectrum (diagrammatic)

to those for a phase modulated sinusoid. Unlike pulse amplitude modulation the sets of spectra occurring at ω_r, $2\omega_r$, etc. are not similar. The magnitude of the sidebands forming the set centred at $k\omega_r$ being determined by the set of functions $J_n(k\phi_0)$. For small ϕ_0, the sidebands are larger and more of them will be in evidence the higher the harmonic of the repetition frequency, as shown in Fig. 6.21. As ϕ_0 becomes larger the spectra will develop in a manner analogous to that for a phase modulated sinusoid. In the case of pulse phase modulation considered in this section there is no spectral term at the modulation frequency p. Hence again in contrast to pulse amplitude modulation a low pass filter will not act as a detector.

6.14 Rowland's Ghosts

The results of the last section can again be interpreted in terms of the diffraction grating. In this case the function $f(x)$ would represent a grating with a periodic displacement of the lines from their true position. Such faults occur as a result of slight imperfections in the ruling engines which can take various forms. An imperfection in the screw gives a periodic error with a period equal to the pitch of the screw. In this case the ghosts are known as Rowland's ghosts. These are more troublesome the higher the order of the spectrum as may be seen from Fig. 6.21 and in the high orders sometimes results in fuzziness of the spectral line. If ϕ_0 is small then we can expand $J_n(k\phi_0)$ as a power series and neglect all but the first term, so that $J_n(k\phi_0) = (k\phi_0)^n/2^n n!$. The amplitude of the ghost nearest to the main line will therefore be proportional to k, and the intensity to k^2, i.e. the intensity of the ghost varies as the square of the order of the spectrum.

6.15 Pulse Phase Modulation—Further Consideration

In section 6.13 we obtained the formula for a phase-modulated train of impulses by replacing $\omega_r t$ by $(\omega_r t + \phi_0 \sin pt)$. Thus if the train of impulses is represented by $f(t)$ we obtained the function $f(t + t_0 \sin pt)$ where $t_0 = \phi_0/\omega_r$. Suppose the impulses occur with a spacing in time of T. Then $f(t)$ is zero except at the points $t = t_i$ where $t_i = \pm nT$ and $n = 0, 1, 2 \ldots$, and at these points an impulse occurs. If now $h(t)$ is given by $f(t + t_0 \sin pt)$ then $h(t)$ is zero everywhere except at the points t_i where $t_i + t_0 \sin pt_i = \pm nT$ and at these points an impulse occurs. Hence it can be seen that the value of the modulating waveform affecting the position of the pulse is its value at time t_i. Thus the position of each pulse is determined by the value of the modulating function at the instant at which that pulse occurs. The spectrum derived in section 6.13 corresponds to a pulse train modulated in this manner.

However the process used in section 6.13 is not justified because the function $h(t)$ representing the phase-modulated train is not in general periodic. It will only be so if ω_r/p is rational. Hence in general we are not justified in assuming a Fourier series type of representation as we did in section 6.13 and we now consider a derivation in which this assumption is not made. In section 4.6 it was shown that a series of N pulses of unit area and width 2τ occurring symmetrically about the origin could be represented by the Fourier integral

$$f(t) = \frac{1}{2\pi}\int_{-\infty}^{\infty} \frac{\sin\omega\tau}{\omega\tau}\frac{\sin\frac{1}{2}N\omega T}{\sin\frac{1}{2}\omega T}\exp(j\omega t)\,d\omega$$

We now obtain $h(t)$ by replacing t by $t+t_0\sin pt$ to give a phase-modulated pulse train of finite duration

$$h(t) = f(t+t_0\sin pt)$$

$$= \frac{1}{2\pi}\int_{-\infty}^{\infty} \frac{\sin\omega\tau}{\omega\tau}\frac{\sin\frac{1}{2}N\omega T}{\sin\frac{1}{2}\omega T}\exp\{j\omega(t+t_0\sin pt)\}\,d\omega$$

Examining the limit $N\to\infty$ by the same procedure as we used in section 4.6 we see that, in the limit, the only contributions to the integral will come from the points where $\sin\frac{1}{2}\omega T = 0$, or the points $\omega = k\omega_r$ where $\omega_r = 2\pi/T$. Hence we can write

$$h(t) = \sum_{k=-\infty}^{\infty} \frac{1}{2\pi}\frac{\sin(k\omega_r\tau)}{k\tau}\exp\{jk\omega_r(t+t_0\sin pt)\}$$

Now letting $\tau\to 0$ to produce impulses we obtain

$$h(t) = \sum_{k=-\infty}^{\infty} \frac{\omega_r}{2\pi}\exp\{jk\omega_r(t+t_0\sin pt)\}$$

$$= \frac{\omega_r}{2\pi}\sum_{k=-\infty}^{\infty}\sum_{n=-\infty}^{\infty} J_n(k\omega_r t_0)\exp\{j(k\omega_r+np)t\}$$

$$\therefore\ h(t) = \frac{\omega_r}{2\pi}\sum_{k=-\infty}^{\infty}\sum_{n=-\infty}^{\infty} J_n(k\phi_0)\exp\{j(k\omega_r+np)t\}$$

6.16 Phase Modulated Pulses of Constant Width

We saw in the last section that the function obtained by replacing t by $(t+t_0 \sin pt)$ produced a series of impulses whose time of occurrence was affected by the value of the modulating function at the instant when the pulse occurred. Similarly if the original function had represented a series of steps instead of impulses the time of occurrence of the step would be affected by the value of the modulating function at the instant the step occurred. Now if the pulses are of finite width we can consider that they are composed of two steps, the first a positive one and the second a negative one occurring at a time 2τ, the width of the pulse, later. Hence from the argument just given if we replace t by $(t+t_0 \sin pt)$ in the expression for pulses of finite width then the position of each edge of the pulses is modified by the value of the modulating function at the instant when the edge occurs. Since in general the modulating waveform will not have the same value at the trailing edge of the pulse as it had at the leading edge of the same pulse, such a pulse train will be width-modulated as well as phase-modulated. If then we require the spectrum of a train of phase-modulated pulses of constant width we must use a process which treats the leading and trailing pulse edges independently.

In section 4.8 we showed that a unit step could be represented by the expression

$$\tfrac{1}{2}+\frac{1}{2\pi}\int_{-\infty}^{\infty}\frac{\exp(j\omega t)}{j\omega}\,d\omega$$

Hence a series of N unit steps occurring at $t=-N'T, -(N'-1)T, \ldots -T, O, T, \ldots (N'-1)T, N'T$, where $N=(2N'+1)$, will be given by

$$\tfrac{1}{2}(2N'+1)+\frac{1}{2\pi}\int_{-\infty}^{\infty}\frac{\exp(j\omega t)}{j\omega}\sum_{n=-N'}^{N'}\exp(jn\omega T)\,d\omega$$

We now phase modulate the position of these steps by replacing t by $(t+t_0 \sin pt)$ giving

$$\tfrac{1}{2}(2N'+1)+\frac{1}{2\pi}\int_{-\infty}^{\infty}\frac{\exp\{j\omega(t+t_0\sin pt)\}}{j\omega}\sum_{n=-N'}^{N'}\exp(jn\omega T)\,d\omega$$

It was shown in section 4.6 that

$$\sum_{n=-N'}^{N'} \exp(jn\omega T) = \frac{\sin \frac{1}{2}N\omega T}{\sin \frac{1}{2}\omega T}, \quad N = 2N'+1$$

Hence the expression above becomes

$$\tfrac{1}{2}N + \frac{1}{2\pi}\int_{-\infty}^{\infty} \frac{\exp\{j\omega(t+t_0 \sin pt)\}}{j\omega} \cdot \frac{\sin \frac{1}{2}N\omega T}{\sin \frac{1}{2}\omega T}\, d\omega$$

To produce a series of phase-modulated pulses of constant width 2τ we shift this waveform τ to the left by writing $(t+\tau)$ in the expression instead of t and subtract from it the same expression shifted τ to the right. Hence we obtain

$$h(t) = \frac{1}{2\pi}\int_{-\infty}^{\infty} \frac{1}{j\omega}[\exp j\omega\{(t+\tau)+t_0 \sin p(t+\tau)\}$$

$$-\exp j\omega\{(t-\tau)+t_0 \sin p(t-\tau)\}] \frac{\sin \frac{1}{2}N\omega T}{\sin \frac{1}{2}\omega T}\, d\omega$$

Now

$$\exp j\omega\{(t+\tau)+t_0 \sin p(t+\tau)\}$$

$$= \exp j\omega t \sum_{n=-\infty}^{\infty} \exp\{j(\omega+np)\tau\} J_n(\omega t_0) \exp(jnpt)$$

and

$$\exp j\omega\{(t-\tau)+t_0 \sin p(t-\tau)\}$$

$$= \exp j\omega t \sum_{n=-\infty}^{\infty} \exp\{-j(\omega+np)\tau\} J_n(\omega t_0) \exp(jnpt)$$

Hence

$$h(t) = \frac{1}{2\pi}\int_{-\infty}^{\infty} \exp(j\omega t) \sum_{n=-\infty}^{\infty} \frac{1}{j\omega}[\exp\{j(\omega+np)\tau\}$$

$$-\exp\{-j(\omega+np)\tau\}] J_n(\omega t_0) \exp(jnpt) \frac{\sin \frac{1}{2}N\omega T}{\sin \frac{1}{2}\omega T}\, d\omega$$

$$= \frac{1}{\pi}\int_{-\infty}^{\infty} \exp(j\omega t) \sum_{n=-\infty}^{\infty} \frac{\sin(\omega+np)\tau}{\omega} J_n(\omega t_0) \frac{\sin \frac{1}{2}N\omega T}{\sin \frac{1}{2}\omega T}\, d\omega$$

Taking the limit $N \to \infty$ as before we get contributions to the integral only at points where $\sin \frac{1}{2}\omega T = 0$ or $\omega = k\omega_r$ where $\omega_r = 2\pi/T$ and

$$h(t) = \frac{1}{\pi} \sum_{k=-\infty}^{\infty} \sum_{n=-\infty}^{\infty} J_n(k\phi_0) \frac{\sin(k\omega_r + np)\tau}{k} \exp j(k\omega_r + np)t$$

which is the expression required in complex form. We can simplify this expression as we did the one in section 6.13. For $k = 0$ we have

$$\frac{1}{\pi}\left[\sum_{n=-\infty}^{\infty} J_n(k\phi_0) \frac{\sin(k\omega_r + np)\tau}{k} \exp(jnpt)\right]_{k=0}$$

$$= \frac{1}{\pi}\left[J_0(0)\frac{\sin k\omega_r\tau}{k} + \frac{J_1(k\phi_0)}{k}\sin(p\tau)\exp(jpt)\right.$$

$$\left. + \frac{J_{-1}(k\phi_0)}{k}\sin(-p\tau)\exp(-jpt)\right]_{k=0}$$

$$= \frac{1}{\pi}[\omega_r\tau + \tfrac{1}{2}\phi_0 \sin p\tau \exp(jpt) + \tfrac{1}{2}\phi_0 \sin p\tau \exp(-jpt)]$$

$$= \frac{1}{\pi}[\omega_r\tau + \phi_0 \sin p\tau \cos pt]$$

For the rest of the series we have

$$\frac{1}{\pi}\sum_{n=-\infty}^{\infty}\left\{\sum_{k=1}^{\infty} + \sum_{k=-1}^{-\infty}\right\} J_n(k\phi_0)\frac{\sin(k\omega_r + np)\tau}{k}\exp j(k\omega_r + np)t$$

$$= \frac{1}{\pi}\sum_{n=-\infty}^{\infty}\sum_{k=1}^{\infty}\left[J_n(k\phi_0)\frac{\sin(k\omega_r + np)\tau}{k}\exp j(k\omega_r + np)t\right.$$

$$\left. + J_n(-k\phi_0)\frac{\sin(-k\omega_r + np)\tau}{-k}\exp j(-k\omega_r + np)t\right]$$

$$= \frac{1}{\pi}\sum_{n=-\infty}^{\infty}\sum_{k=1}^{\infty}\left[J_n(k\phi_0)\frac{\sin(k\omega_r + np)\tau}{k}\exp j(k\omega_r + np)t\right.$$

$$\left. + J_{-n}(-k\phi_0)\frac{\sin(-k\omega_r - np)\tau}{-k}\exp j(-k\omega_r - np)t\right]$$

$$= \frac{1}{\pi} \sum_{n=-\infty}^{\infty} \sum_{k=1}^{\infty} \left[J_n(k\phi_0) \frac{\sin(k\omega_r + np)\tau}{k} \{\exp j(k\omega_r + np)t + \exp j(-k\omega_r - np)t\} \right]$$

$$= \frac{2}{\pi} \sum_{n=-\infty}^{\infty} \sum_{k=1}^{\infty} J_n(k\phi_0) \frac{\sin(k\omega_r + np)\tau}{k} \cos(k\omega_r + np)t$$

Hence for the complete expression we have

$$h(t) = \frac{1}{\pi}\left[\omega_r \tau + \phi_0 \sin p\tau \cos pt + 2 \sum_{n=-\infty}^{\infty} \sum_{k=1}^{\infty} J_n(k\phi_0) \frac{\sin(k\omega_r + np)\tau}{k} \cos(k\omega_r + np)t\right]$$

We see that there is now a term at the modulation frequency, p, i.e. $\phi_0 \sin p\tau \cos pt$. However its magnitude, $\phi_0 \sin p\tau$, is a function of the modulation frequency and hence for general modulating waveforms the low frequency terms do not represent the original modulating waveform. A low pass filter would not therefore be a suitable detector.

The expression $\sum_{n=-\infty}^{\infty} J_n(k\phi_0)\cos(k\omega_r + np)t$ represents a phase modulation of the carrier $\cos k\omega_r t$ by the function $\cos pt$ with a phase excursion of $k\phi_0$ (see section 6.7). A phase detector centred at $k\omega_r$ would produce the original waveform from this set of terms. However in the expression above this term is modified by the factor $\sin(k\omega_r + np)\tau$ which again is a function of p, the modulation frequency, so again we cannot obtain an undistorted output.

It is common practice to detect pulse phase modulation by first converting it into pulse width modulation.

6.17 Pulse Width Modulation

In pulse width modulation the time duration of the pulses is modified by the modulating waveform. We shall assume the leading edges of the pulses are independent of the modulation and therefore occur at constant time intervals and that the position of the trailing edge is modified in accordance with the modulation waveform. It can be seen that on this assumption it is not possible to construct a symmetrical waveform. We shall suppose that a positive going edge (i.e. a leading edge) occurs at $t = 0$. Constructing the pulses by a process similar to that used above we have for the series of steps representing the leading edges

$$\tfrac{1}{2}N + \frac{1}{2\pi j}\int_{-\infty}^{\infty} \frac{\exp(j\omega t)}{\omega}\frac{\sin\frac{1}{2}N\omega T}{\sin\frac{1}{2}\omega T}\,d\omega$$

If the mean length of the pulse is τ then the negative steps must first be displaced in time by $+\tau$ which is obtained by replacing t by $(t-\tau)$. We then wish to modulate their position (or phase) by $\tau_0 \sin pt$ which we do by replacing t by $(t+\tau_0 \sin pt)$. The maximum pulse width will be $(\tau+\tau_0)$, and the minimum $(\tau-\tau_0)$. The depth of modulation will be τ_0/τ. Carrying out the two operations above we have for the required series of negative steps

$$-\tfrac{1}{2}N - \frac{1}{2\pi j}\int_{-\infty}^{\infty} \frac{\exp\{j\omega(t-\tau+\tau_0\sin pt)\}}{\omega}\frac{\sin\frac{1}{2}N\omega T}{\sin\frac{1}{2}\omega T}\,d\omega$$

Hence for the complete waveform we have

$$h(t) = \frac{1}{2\pi j}\int_{-\infty}^{\infty} \frac{1}{\omega}[\exp(j\omega t) - \exp\{j\omega(t-\tau+\tau_0\sin pt)\}]\frac{\sin\frac{1}{2}N\omega T}{\sin\frac{1}{2}\omega T}\,d\omega$$

taking the limit $N\to\infty$ as before we have

$$h(t) = \frac{1}{2\pi j}\sum_{k=-\infty}^{\infty}\left[\frac{\exp(jk\omega_r t)}{k} - \frac{\exp(jk\omega_r t)\exp(-jk\omega_r \tau)\exp(jk\omega_r \tau_0 \sin pt)}{k}\right]$$

$$= \frac{1}{2\pi j}\sum_{k=-\infty}^{\infty}\left[\frac{\exp(jk\omega_r t)}{k} - \frac{\exp(jk\omega_r t)}{k}\exp(-jk\omega_r \tau)\sum_{n=-\infty}^{\infty} J_n(k\omega_r \tau_0)\exp(jnpt)\right]$$

which is the required expression in complex form. A sketch of the waveform represented by this expression is shown in Fig. 6.22. Simplifying as above we consider first $k = 0$ giving

$$\frac{1}{2\pi j}\left[\frac{\exp(jk\omega_r t)}{k} - \frac{\exp(jk\omega_r t)}{k}\exp(-jk\omega_r \tau)\{J_0(0)\right.$$

$$\left. + J_1(k\omega_r \tau_0)\exp(jpt) + J_{-1}(k\omega_r \tau_0)\exp(-jpt)\}\right]_{k=0}$$

$$= \frac{1}{2\pi j}\left[\frac{\exp(jk\omega_r t)}{k} - \frac{\exp(jk\omega_r t)}{k}\exp(-jk\omega_r \tau)\right.$$

$$- \frac{\exp(jk\omega_r t)}{k}\exp(-jk\omega_r \tau)\{\tfrac{1}{2}k\omega_r \tau_0 \exp(jpt)$$

$$\left. - \tfrac{1}{2}k\omega_r \tau_0 \exp(-jpt)\}\right]_{k=0}$$

$$= \frac{1}{2\pi j}\left[\frac{\exp\{jk\omega_r(t-\frac{1}{2}\tau)\}}{k}\{\exp(j\tfrac{1}{2}k\omega_r \tau) - \exp(-j\tfrac{1}{2}k\omega_r \tau)\}\right.$$

$$\left. - \tfrac{1}{2}\omega_r \tau_0 \exp\{jk\omega_r(t-\tau)\}\{\exp(jpt) - \exp(-jpt)\}\right]_{k=0}$$

$$= \frac{1}{\pi}\left[\frac{\sin(\frac{1}{2}k\omega_r \tau)}{k} - \tfrac{1}{2}\omega_r \tau_0 \sin pt\right]_{k=0} = \frac{1}{2\pi}[\omega_r \tau - \omega_r \tau_0 \sin pt]$$

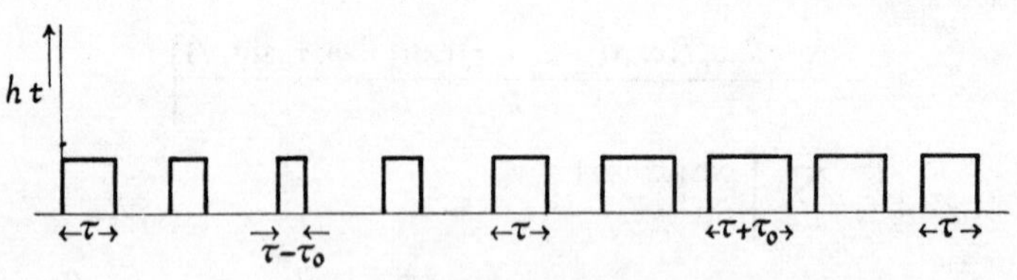

FIG. 6.22
Pulse width modulated signal

For the rest of the series ($k \neq 0$) we evaluate the two terms separately. The first gives

$$\frac{1}{2\pi j}\left\{\sum_{k=1}^{\infty}+\sum_{k=-1}^{-\infty}\right\}\frac{\exp(jk\omega_r t)}{k}=\frac{1}{2\pi j}\sum_{k=1}^{\infty}\frac{\exp(jk\omega_r t)}{k}-\frac{\exp(-jk\omega_r t)}{k}$$

$$=\frac{1}{\pi}\sum_{k=1}^{\infty}\frac{\sin(k\omega_r t)}{k}$$

and the second gives

$$\frac{-1}{2\pi j}\left\{\sum_{k=1}^{\infty}+\sum_{k=-1}^{-\infty}\right\}\frac{\exp\{jk\omega_r(t-\tau)\}}{k}\sum_{n=-\infty}^{\infty}J_n(k\omega_r\tau_0)\exp(jnpt)$$

$$=\frac{-1}{2\pi j}\sum_{k=1}^{\infty}\sum_{n=-\infty}^{\infty}\left[\frac{\exp\{jk\omega_r(t-\tau)\}}{k}J_n(k\omega_r\tau_0)\exp(jnpt)\right.$$

$$\left.-\frac{\exp\{-jk\omega_r(t-\tau)\}}{k}J_n(-k\omega_r\tau_0)\exp(jnpt)\right]$$

$$=\frac{-1}{2\pi j}\sum_{k=1}^{\infty}\sum_{n=-\infty}^{\infty}\left[\frac{\exp\{jk\omega_r(t-\tau)\}}{k}J_n(k\omega_r\tau_0)\exp(jnpt)\right.$$

$$\left.-\frac{\exp\{-jk\omega_r(t-\tau)\}}{k}J_{-n}(-k\omega_r\tau_0)\exp(-jnpt)\right]$$

$$=\frac{-1}{2\pi j}\sum_{k=1}^{\infty}\sum_{n=-\infty}^{\infty}\left[\frac{1}{k}J_n(k\omega_r\tau_0)\{\exp j[(k\omega_r+np)t-k\omega_r\tau]\right.$$

$$\left.-\exp[-j(k\omega_r+np)t+jk\omega_r\tau]\}\right]$$

$$=-\frac{1}{\pi}\sum_{k=1}^{\infty}\sum_{n=-\infty}^{\infty}\frac{1}{k}J_n(k\omega_r\tau_0)\sin\{(k\omega_r+np)t-k\omega_r\tau\}$$

Hence we have finally

$$h(t) = \frac{1}{2\pi}\left[\omega_r(\tau-\tau_0\sin pt)+2\sum_{k=1}^{\infty}\frac{1}{k}\sin(k\omega_r t)\right.$$

$$\left.-2\sum_{k=1}^{\infty}\sum_{n=-\infty}^{\infty}\frac{1}{k}J_n(k\omega_r\tau_0)\sin\{(k\omega_r+np)t-k\omega_r\tau\}\right]$$

This is a more awkward looking expression than those we have obtained previously because the function is not symmetrical about the origin. It is worth while making a check to see that it reduces to the correct form in the absence of modulation, i.e. $\tau_0 = 0$. In this case we have

$$h(t) = \frac{1}{2\pi}\left[\omega_r\tau+2\sum_{k=1}^{\infty}\frac{1}{k}\sin(k\omega_r t)-2\sum_{k=1}^{\infty}\frac{1}{k}\sin k\omega_r(t-\tau)\right]$$

$$= \frac{1}{2\pi}\left[\omega_r\tau+2\sum_{k=1}^{\infty}\frac{1}{k}\{\sin(k\omega_r t)-\sin k\omega_r(t-\tau)\}\right]$$

$$= \frac{1}{2\pi}\left[\omega_r\tau+4\sum_{k=1}^{\infty}\frac{1}{k}\sin(\tfrac{1}{2}k\omega_r\tau)\cos k\omega_r(t-\tfrac{1}{2}\tau)\right]$$

We can check this from the formula developed in section 2.6. From these results we see that for pulses of duration τ (note that in section 2.6 a duration of 2τ was used) occurring symmetrically about the origin we have

$$a_0 = \frac{2\tau}{T} = \frac{2\omega_r\tau}{2\pi} = \frac{\omega_r\tau}{\pi}$$

and

$$a_k = \frac{2}{\pi k}\sin(\tfrac{1}{2}k\omega_r\tau)$$

so we can write

$$f(t) = \tfrac{1}{2}a_0 + \sum_{k=1}^{\infty} a_k \cos k\omega_r t$$

$$= \frac{\omega_r \tau}{2\pi} + \sum_{k=1}^{\infty} \frac{2}{\pi k} \sin(\tfrac{1}{2}k\omega_r \tau)\cos(k\omega_r t)$$

$$= \frac{1}{2\pi}\left[\omega_r \tau + 4\sum_{k=1}^{\infty} \frac{1}{k} \sin(\tfrac{1}{2}k\omega_r \tau)\cos(k\omega_r t)\right]$$

Now to obtain the same function as above we must shift this pulse train so that a leading edge occurs at $t = 0$ instead of $t = -\frac{1}{2}\tau$. Hence replacing t by $(t-\frac{1}{2}\tau)$ we have

$$f(t) = \frac{1}{2\pi}\left[\omega_r \tau + 4\sum_{k=1}^{\infty} \frac{1}{k} \sin(\tfrac{1}{2}k\omega_r \tau)\cos k\omega_r(t-\tfrac{1}{2}\tau)\right]$$

which checks the result above. It can also be seen in a general way from Fig. 6.22 that the component $\sin pt$ will be negative since the coefficient will be given by the formula

$$b_1 = \frac{2}{T}\int_{-T/2}^{T/2} f(t)\sin pt\, dt$$

and it can be seen that the wide pulses occur where the sine function is negative and hence the negative contributions to the integral will be the larger. This is also in agreement with the expression obtained.

In the expression for the pulse width modulated wave there is a term at the modulation frequency $\sin pt$ whose magnitude is independent of p and proportional to τ_0 the depth of modulation. The undistorted modulation waveform can therefore be obtained from this and a low pass filter will act as a detector.

6.18 Abbé Theory of Image Formation

In the early part of the chapter we developed the idea of Fourier pairs in which the functions were functions of space variables instead of time variables and the similarity between diffraction patterns and the frequency spectra of time functions was demonstrated. The Abbé theory of image formation is closely parallel to the case of filtering which we discussed in section 5.2. There we started with a function of time which we transformed into its spectrum, the spectrum was modified by the filter and the output time function was the inverse transform of the resulting frequency function. The Abbé theory of image formation states that for a truthful image of a structure to be formed by a lens the aperture must be wide enough to transmit the whole of the diffraction pattern produced by the structure; if only a portion is transmitted then the image produced corresponds to an

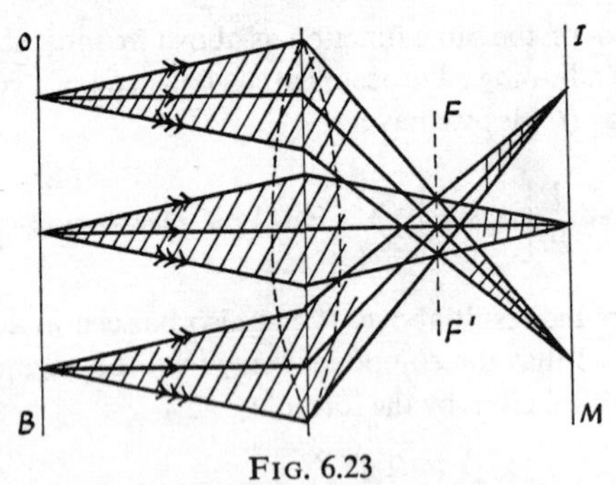

FIG. 6.23
The formation of an image by a lens

object whose complete diffraction pattern consists of the portion transmitted. Consider a one-dimensional structure represented by $f(x)$. Then as we have seen the diffraction pattern is $F(K)$ where $K = (2\pi \sin\theta)/\lambda$. If the aperture of the lens is limited then light rays within only a finite range of θ are transmitted and hence $F(K)$ is modified to $F_1(K)$ say, and the image becomes $f_1(x)$ the inverse transform.

Consider for example a grating in the plane OB (Fig. 6.23) illuminated by monochromatic light. Bundles of rays emanating from three of the slits are shown in the figure. The rays marked with one arrow are parallel to the axis and will therefore come to a focus at the point F_1 the principal focus of the lens. Suppose the rays marked by a double arrow are in the direction of the first spectrum. These are all parallel to each other and therefore come to a focus in the focal plane FF' at some point F_2. Similarly for the triple arrowed rays in the direction of the first spectrum on the other side of the axis. Thus the spectra are formed in the focal plane of the lens. The bundles of rays from the points in the object plane come to a focus in the image plane IM and form an image. If however a stop were placed in the focal plane so that only the first order spectra were transmitted then the image would not be a true image of the grating in the OB plane but would correspond to the inverse transform of the spectra transmitted. In this case where only the first order are transmitted a sinusoidal variation of intensity would be formed in the IM plane. The effects of passing only a limited number of spectra can be demonstrated experimentally. These results are of practical significance in microscopy where on occasions the object may differ considerably from the image due to modification of the spectra by the aperture of the instrument so that care is needed in interpreting the results.

REFERENCE

ARGUIMBAU, L. B. and STUART, R. D. *Frequency Modulation*, Methuen (1956)

APPENDIX

DIFFERENTIATION OF A DEFINITE INTEGRAL WITH RESPECT TO ITS UPPER LIMIT

We wish to evaluate

$$\frac{d}{dt}\int_0^{at} g(x)\,dx$$

where a is a constant.

Let $$h(x) = \int g(x)\,dx$$

so that

$$\frac{d}{dx}h(x) = g(x)$$

Then

$$\int_0^{at} g(x)\,dx = h(at) - h(0)$$

$$\therefore \frac{d}{dt}\int_0^{at} g(x)\,dx = \frac{d}{dt}h(at)$$

$$\therefore \frac{d}{dt}\int_0^{at} g(x)\,dx = ag(at)$$

Index